EMBRACING EARTH

EMBRACING EARTH

Catholic Approaches to Ecology

Edited by

Albert J. LaChance
John E. Carroll

ORBIS BOOKS

Maryknoll, New York 10545

The Catholic Foreign Mission Society of America (Maryknoll) recruits and trains people for overseas missionary service. Through Orbis Books, Maryknoll aims to foster the international dialogue that is essential to mission. The books published, however, reflect the opinions of their authors and are not meant to represent the official position of the society.

Library of Congress Cataloging-in-Publication Data

Embracing earth : Catholic approaches to ecology / edited by Albert J.
LaChance, John E. Carroll.
        p.   cm.
    Includes bibliographical references
    ISBN 0-88344-966-8
    1. Human ecology—Religious aspects—Catholic Church   2. Catholic
Church—Doctrines.   I. LaChance, Albert J.   II. Carroll, John E.
(John Edward), 1944–   .
BX1795.H82E42   1994
261.8'362'08822—dc20   94-31670
                                                                CIP

Dedicated to

Rebecca LaChance
Kateri LaChance
Abby Carroll
Cozette Carroll
And to all children and young of all species

All things by immortal power
Near or far . . .
To each other linked are,
That thou canst not stir a flower
Without troubling of a star.

—Francis Thompson
"The Mistress of Vision"

Contents

Ecology and the Future of Catholicism xi
A Statement of the Problem
    Thomas Berry, C.P.

Foreword xiii
    Miriam Therese MacGillis, O. P.

Introduction xvii
    Elucidating Fundamental Values   *xvii*
    John E. Carroll
    Elucidating Catholic Values   *xxi*
    Albert J. LaChance

1.   Earthly Offerings 1
    Sacrifice and the Land in Ancient Israel and Early Christianity
    Frederick G. Levine

2.   God, the Cosmos, and Culture 15
    Albert J. LaChance

3.   Christ the Ecologist 30
    John E. Carroll

4.   Open to Life—and to Death 35
    The Church on Population Issues
    David S. Toolan, S.J.

5.   A Revolution in Human Ecology 47
    Mary Rosera Joyce

6.   The Theotokos Project 53
    Beatrice Bruteau

7.   A Loaves and Fishes View of Productivity 76
    Richard C. Haas

8.   Appropriate Technology and Healing the Earth 96
    Albert J. Fritsch, S.J.

9.   Eating the Body of the Lord 115
    Eucharist and Community-Supported Farming
    Marc Boucher-Colbert

10. Christianity and the Creation 129
     A Franciscan Speaks to Franciscans
     Richard Rohr, O.F.M.

11. Fruit of the Earth, Fruit of the Vine 156
     Charles Cummings, O.C.S.O.

12. Ecological Resources in the Benedictine Rule 163
     Terrence G. Kardong, O.S.B.

13. Conversation with the Cosmic Christ 174
     The Spiritual Exercises *from an Ecological Perspective*
     William J. Wood, S.J.

14. Gaia—Samsara—Narnia 192
     Tessa Bielecki, O.C.D.

15. Verbal Pollution 203
     William McNamara, O.C.D.

16. An Eco-Prophetic Parish? 214
     Paula González, S.C.

17. Was St. Francis a Deep Ecologist? 225
     Keith Warner, O.F.M.

18. Choose Life 241
     Ascetic Theology, History, and Ecology
     David M. Sherman

19. Concluding Reflections 255
     Toward a Second Axial Age
     Wayne Teasdale

A Prayer for Animals 276

Contributors 277

Ecology and the Future of Catholicism

A Statement of the Problem

THOMAS BERRY, C.P.

The future of the Catholic church in America, in my view, will depend above all on its capacity to assume a religious responsibility for the fate of the earth. For ecclesiastical authorities to be so negligent, indifferent, or unaware of the imperiled status of the planet in all its living and nonliving components can be considered as a monumental failure. In order of magnitude, this goes far beyond any other failure for which the church may be responsible.

The devastation that is presently taking place through the industrial-technological exploitation of the earth is already so great that future generations are presently condemned not only to live amid the ruins of the infrastructures of industrial society but also amid the ruins of the natural world itself. This can be said apart from any consideration of nuclear war. So far, church authorities, religious orders, the Catholic universities and seminaries, priests, and people have shown an amazing insensitivity to this most urgent of all issues confronting the human community. My question is: After we burn our lifeboat, how will we stay afloat? What will then be the need of religion, Christianity, or the church?

Presently the church has a unique opportunity to place its vast authority, its energies, its educational resources, its spiritual disciplines in a creative context, one that can assist in renewing the earth as a bio-spiritual planet. If this is not done immediately, then by the end of this century an overwhelming amount of damage will be done, an immense number of living species will be irreparably lost for all future generations. Only by assuming its religious responsibilities for the fate of the earth can the church regain any

Thomas Berry wrote this essay in response to an inquiry from the sociology department of a Catholic university in November 1982.

effective status either in the human community or in the earth process.

Hundreds of groups proceeding on a program of bio-cultural regionalism are already engaged in reinhabiting the earth within the context of the ever-renewing sequences of life in the natural world. Reinhabiting the North American continent on a basis of the mutual enhancement of the human and the natural is the task to which the church in America should be dedicated. Even our concerns for international peace and for social justice can be realized only within this context. Peace among the nations will come only through peace with the earth. A truly functional human society can exist only within the context of a functional earth society.

The renewal of religion in the future will depend on our appreciation of the natural world as the locus for the meeting of the divine and the human. The universe itself is the primary divine revelation. The splendor and the beauty of the natural world in all its variety must be preserved if any worthy idea of the divine is to survive in the human community.

Foreword

MIRIAM THERESE MacGILLIS, O.P.

During the month of August 1974, I caught a late afternoon bus out of New York City and rode through the night, reaching Bar Harbor, Maine, at dawn. From there I took a long ferry ride for the last leg of a journey which ended in a primitive cabin deep in the remote woodlands of Nova Scotia. Nova Nada Spiritual Life Center had been founded by William McNamara, O.C.D., to welcome pilgrims like myself who were longing for the silence and solitude that the wilderness alone can provide.

I was thirty-four years old. Still dressed in a full-length Dominican habit, coif and veil, I was a strange-looking passenger on that night bus. I had received approval to follow this unquenchable thirst to go as far apart as I could from everything I knew and spend two weeks in solitude in the Canadian woods. What I did not realize at the time was that I was in search of the survival of my soul-life. It had been sixteen years since I had left my family to enter religious life, and those years spent in urban and suburban settings had drastically cut me off from the woods, rivers, and wild places in which my summers had been rooted.

When I was seven years old, my parents brought us four children to the undeveloped woodlands of the Musconetcong watershed of northwestern New Jersey. We built a small cabin over the summers of my childhood, living most of those years without water or electricity. The nearest store or phone was an hour's walk away. The trees and birds, the snakes, marshes, and rivers, the night sounds and the physical challenges had shaped my soul. I lacked the concepts or language for identifying the vast inner longings that had gone unreprieved in the years since I had left that life behind. But at thirty-four years of age, I was in an intense stage of transition, and this trip to Nova Scotia was like life or death.

Raised in the '40s and '50s in a totally "Catholic" world, I grew up innocent and secure in my own identity and purpose, in an unquestioned clarity about the meaning of everything, and in an absolute faith in my church and country.

By 1974, my faith had been shaken to its foundation. The racial and economic inequities within America had erupted throughout the '60s. The hidden, shadow-side of the Vietnam War had come to light. A rising awareness of global power-brokers, of militarism and secret corruption, of untold human suffering around the world had pierced the naivete of my youth. But it was the slow, difficult discipline of the renewal of religious life, imposed on women religious in the aftermath of Vatican II, which was honing a critical edge to my thinking and empowering me to ask questions about everything.

Like most religious women, I was carried by the wave of that renewal process that we accepted in obedience. It moved us through discernment, social and structural analysis, a passionate search for the "signs of the times," and to listening to the unwritten histories of the poor and of women. These explorations sharpened the critical edge of our consciousness. We were penetrating to the roots of the human crisis, the American crisis, the patriarchal crisis, and what was emerging was an expansion of our religious commitment, from the small cultural boundaries which had defined it formerly, to a vast, unquestioned commitment to life wherever it was threatened.

Trained as an artist and teacher, I had recently left both professions and had apprenticed myself in the arena of Catholic social teaching on justice and peace. I was in a profound transition, inwardly and outwardly, and the river flowing beneath all this searching was an intense longing to be immersed in the natural world. When I returned from the Canadian woods I knew that nature was as essential to my life as air and food. But I had no way to express that knowledge.

It was in the early spring of 1977 that form and words were given to the unnamed tensions of my life. During a conference entitled Christian Voices on World Order sponsored by Global Education Associates, Thomas Berry delivered a paper entitled "Contemplation and World Order." At the time he was an associate professor of religion at Fordham University. He was inviting this group of Christian theologians to focus, not so much on contemplation, but on the world order that was being contemplated. Using concepts and language utterly unfamiliar to me, he was challenging us to contemplate the full, spiritual dimensions that were revealed in the scientific explanation of the origin and nature of the universe. He held up this "sacred story" as the greatest religious revolution since the origins of religion. He spoke of the human as that "being in whom the earth and universe had become self-reflexively conscious of itself."

My single most vivid memory of the quiet, undramatic reading of his paper is that I could not follow a word of it. I couldn't fill in the spaces of his thought. Afterward I could not even articulate what he had said. But my whole being was literally shaking, and at some deep, prerational level, I knew I had just heard one of the most significant messages I would ever receive.

My body and my psyche were resonating with words and understandings that confirmed the deep, inner longings of my spirit. I realized I was neither crazy nor abnormal because of my need to be close to the natural world. I had just received a scientific verification that my physical and spiritual longings were in fact intact with the universe which was my deeper self; I had come home to myself as *earthself.*

At the same time Berry was challenging us to contemplate the destruction being visited on the entire web of life, which was my own life. He was imploring us as Christian thinkers to direct our contemplative gaze toward the universe itself and realize that science had, in fact, discovered the "soul of the world." This soul was being diminished most severely by those, committed to the biblical worldview, whose technologies reflected a dichotomy between spirit and matter. This, he suggested, was essential to addressing the crisis of the earth.

The collection of stories in this book has been drawn from a variety of Catholics who are also responding to the planetary crises. They give varied personal witness to the deep awakenings happening within themselves and in the lives of many Catholics.

We are all children of this explosive century. Regardless of our urban or rural roots, we began our lives in a religious culture shaped by the astounding promises of the industrial age. The world of unlimited technology, prosperity, and individual liberty was an unquestioned endowment of the past. Whether our present capacity to critique that assumption is awakened through a mystical, scriptural, biological, cosmological, or philosophical impulse, we have all acknowledged the dire consequences that our human species is visiting on the entire earth and all its communities of life.

What holds us together as ecological Catholics is an unbroken continuity with the inner psychic world shaped by the countless individual people who have lived out the ancient faith development of Hebrew and Christian traditions. The sacredness of life, the fidelity of the *Divine-Holy-Nameless-One*, whose presence is in story and journey with life itself, the emergence of law and covenant, of righteousness in relationship, and ultimately of sacrificial love and community—all these are among the rich threads that have

woven the seamless garment of the sacramental life of Catholics.

What marks the awakening we experience today is the special
clarity of awareness that this seamless garment is the entire fabric
of earth and universe, not only its human aspect. We have come to
understand the earth and universe as one with itself, and thus the
categories in which the mysteries of the Christian life arose are
being challenged to expand beyond the historic forms which birthed
them into the larger cosmic forms awaiting them.

There need be no denial of the past or of tradition by Catholics
entering into the ecological stage of Christian evolution. Rather
there must be fidelity and hope, expressed in a conversion more
challenging and demanding than ever, because it is toward the
total community of life in a still-expanding universe.

May these reflections be a bridge that brings together the world
of science and the world of religion into a cohesive story of unity
longed for by the children of this age. May the mind and body of
the earth become one in each of us, so that the unity longed for in
prophecy and dream can be celebrated in the universal catholic
community of earth.

Introduction

We wish to bring into relief several singularities of the book you hold. We editors, John Carroll and Albert LaChance, are conscious of running against a prevalent ecumenical grain by asking a group comprised almost only of Catholics (Frederick Levine, a Jew, is the single exception) to put their minds to the task of Catholicism in ecology and ecology in Catholicism.

We do so *without* denigrating Christian ecumenism and *with* a sense that in the great pantheon of traditions attempting to make the message concrete, witness the truth, and incarnate the person of Jesus the Christ individual traditions like Catholicism need moments when they converse within the family circle about ways to renew their life together. *Embracing Earth* is such a book. We invite our brothers and sisters in other Christian traditions, as well as in world religious traditions, to listen in on our conversation.

There is a great variety here. It runs from the highly mystical to the rigorously analytical to the curmudgeonly caustic. We believe there is plenty to inspire, to confound, and to anger persons from every ideological and theological shade. But this is also one of the paradoxes of Catholicism we hope our volume makes visible. We are one but diverse.

May the many members make one body!

Elucidating Fundamental Values

JOHN E. CARROLL

Catholics believe in "God, the Father Almighty, *Creator of heaven and earth.*" They believe in the created order of heaven. They believe in the created order of earth. Catholics believe in the biblical dictum, "The earth is the Lord's, and the fullness thereof." This being so, Catholic belief in God must include belief in the integrity of the earth, God's creation. Those who do not so believe can hardly consider themselves Catholic.

In spite of a degree of uncertainty and contradiction, which is ever present on any scientific question, the evidence given to us by the highest principles and the best practice of our contemporary Western science tells us, clearly and vigorously, of the decline in the condition for the sustainability of life on earth, including the conditions necessary for the sustenance of our own species, *Homo sapiens*. More than simply destroying the ecosystem itself, the forests, the seas, the air, the water, the soil, the animal and plant species, and natural communities of our planet, it is now clear that we are tearing asunder the ecological fabric, the ecological pattern, the ecological system necessary to our own survival and, as well, that of a great diversity of species. Our civilization and its value system are now rendering vast areas of the earth uninhabitable by our species and driving countless other species to extinction.

We now see and hear daily about pollution:

- of ground water aquifers, and of their depletion as well as contamination;
- of surface waters of all kinds, fresh and salt;
- of air quality, local, regional, even hemispheric and global.

We hear on all sides about:

- the tremendous rate of depletion and erosion of soils, the very source of our food and so much else, and of the not unrelated creeping desertification worldwide;
- forest destruction, temperate as well as tropical, of ancient old-growth as well as the not-so-old, and of the conversion of a normally renewable forest resource to a nonrenewable resource through the destruction of the conditions necessary for tree and forest renewal and regeneration;
- species' destruction, plant and animal, through extinction, increasing at nearly geometrical rates and for all time. This involves the loss forever of the unique and complex genetic heritage, the DNA strand, of whole categories of plant and animal life, whose value to us humans as well as to the natural process we can hardly imagine, never mind fathom;
- our increasing ability to alter earth's climate, the very bio- and geo-chemical system of our planet, speeding evolution and change to such a degree that no species, including our own, could conceivably keep pace. Our newfound ability to destroy our planet's protection against extra-terrestrial galactic elements can readily and easily terminate our existence through the destruction of the stratospheric ozone layer.

This list, this litany, can go on and on and is increasingly familiar to all of us. We can take no comfort from the fact that there has been, continues to be, and always will be contradiction and controversy within the science which is telling us these things. We will never be without such contradiction. But the signs, the weight of the evidence, clearly indicate to all but perhaps the most ignorantly obstinate that these things are now occurring. If we consider ourselves to be modern, rational, scientific people, we have no grounds, no rationale, for rejecting the message. If we do believe otherwise, as we are certainly entitled to do, then it is important to be honest with ourselves in admitting we have come to such a conclusion based on faith and in contradiction to prevailing scientific evidence. Indeed, there is nothing wrong with reaching such a conclusion on grounds of faith, as long as we are honest enough to recognize that, in this instance, we are clearly rejecting overwhelming scientific evidence and therefore rationality in doing so.

And what does our science and hence our rationality tell us? They tell us that our usage of and our demand for the world's ecosystems have put us beyond the planet's limits, that we are "drawing down the planet's supporting resources and overflowing its pollution sinks," its pollution-absorption capacity.[1]

If we are seriously to consider the implications of our ecological reality, it is wise to have some sense of the basic rules or principles of ecology.

The science of ecology, we are told by American ecologist Eugene B. Odum, is the study of organisms, including ourselves, "at home."[2] It is the study of the relation of organisms or groups of organisms to their environment; the science of the interrelations, interconnections, interdependencies of organisms to their environment. It is the study of the very structure and function of nature itself (including ourselves), of natural process, of the fabric, the pattern, of life in the cosmos, the organic and the inorganic combined. It addresses the inseparable interrelationship and interaction of the living (biotic) and nonliving (abiotic) environment. And it accepts as reality that individual organisms (again including ourselves) not only adapt to the physical environment but, by their concerted action in ecosystems, also adapt the geochemical (physical) environment to their biological needs.

Ecology speaks of a system in balance, of a self-regulating and self-maintaining system. Built-in limiting factors maintain this homeostatic or balanced system. This system or ecosystem is not a thing, an object, but rather an ongoing process taking place through a complex system of interrelationships both between and among organisms and between organisms and their physical environment. Thus, there is a center of balance or homeostasis around which the states of the ecosystem fluctuate and which defines its self-identity. This center can be impaired, damaged, degraded; its self-identity can be ruptured. Available scientific evidence (and numerous physical signs) indicates this is just what is happening in these last years of the twentieth century. Our species, *Homo sapiens*, represents the cause. If "the earth is the Lord's and the fullness thereof," and, indeed, our home, there will be a price to pay.

Barry Commoner, in his classic work *The Closing Circle*, has provided us with the four laws of ecology:

1. Everything is connected to everything else.
2. Everything must go somewhere.
3. Nature knows best.
4. There is no such thing as a free lunch.[3]

While we give lip service to these laws, especially the first (which all ecological scientists claim to believe), by our actions we daily demonstrate that we simply do not believe these laws. These laws, if followed, would significantly constrain our present way of life in Western society and bring about such fundamental change in the way we live that we would likely not recognize the "humanscape" that would result. Indeed, the first alone, if believed, would bring about change in human behavior of the most fundamental kind, for it would mean we have no escape from relationship, from involvement, from responsibility.

Our modern Western rational science tells us that ecology is valid and that these laws of ecology are valid. Behaving irrationally, however, we choose not to believe them; such belief would cause change we are not willing to undergo. Hence we choose, unscientifically and irrationally, to ignore them. The planet, the declining global ecosystem we witness, is the result of that choice.

We must conclude from the results of this choice we have made that the environmental question, the ecological question, is not fundamentally a question of science and technology, even though science and technology play a role in its resolution; that the environmental question is not fundamentally a question of economics, even though economics plays a role in its resolution; that the environmental question is not fundamentally a question of politics, of political science, of governance, of diplomacy, or of international affairs, even though all of these disciplines play a role. If not these, then what? The environmental question, the ecological question, the energy question, the natural resource question, the agriculture question, is fundamentally a question of ethics and values. The question is a philosophical question, even a theological question, a religious question, a question of values, both secular and spiritual, and the question can only be tackled on this level.

What to do? A fundamental values change is necessary—as explicitly identified in my chapter in this volume, "Christ the Ecologist," and at least implicitly identified in all the other chapters of this book. Along with such change, permanent and significant behavioral lifestyle alterations will occur. Living lightly, fulfilling life's needs in ways that are not consumer driven, virtual elimination of waste and even of the concept of waste will result, as will the conduct of life within ecological laws. With such conduct, our ecological challenges will come to be resolved.

Elucidating Catholic Values

ALBERT J. LaCHANCE

I am a Catholic. The word itself comes from the Greek words *kata* and *holos*. *Holos* is the root for our English word "whole" and the "hol" in holistic. So why is it that *catholic* and *holistic* seem so often to be understood as antonyms?

I am a catholic in that my concern is for the whole of creation. I am also Catholic in the other sense. I believe that the pope is Christ's vicar on earth. I believe that he has the role of teaching Christ's revelation for the world with divine authority. Why would I belong to a religion that I did not believe was divine and revelatory?

I also believe in the basic truths of the Catholic church—the incarnation of God in Jesus, the virgin birth of Jesus, redemption, forgiveness of sin, bodily resurrection. My job as believer is to accept revealed truth as truth, and then to struggle to understand why it is truth. I have done that, but I can also understand why this truth can seem so strange, why Jesus is still a stumbling block for non-Christians and absurdity to those who depend on reason.

I believe in the sacraments. Water reveals its divine powers in baptism. The fire of the Spirit is revealed in confirmation. The holiness of food is revealed in the eucharist. The sacredness of sex is revealed in marriage. When DNA races to embrace DNA, God's own life is revealed. My marriage has been lived within the truth I see in Pope Paul VI's encyclical *Humanae Vitae*, without ignoring the problem of overpopulation or denying physical desire.

In reconciliation, God's forgiveness is revealed. In anointing, the potential of healing is revealed in oils pressed from earth's vegetables. In orders, the self-emptying of God in Christ Jesus is revealed in those who truly surrender to a life of poverty and service. The sacraments are windows in the universe through which the very life of the Creator is revealed. Why walk away from miracles freely given?

I first studied Catholic theology with the Benedictine monks of St. Anselm College in Manchester, New Hampshire, under monks like Peter Guerin and Placidus Riley, as well as under laymen such as James McGhee. I wrestled with the theological and philosophical foundations of Roman Catholicism, and in 1984 I went to California, where I studied with Matthew Fox and Brian Swimme. I spent five years working with Thomas Berry. While it is common to accuse Matthew Fox and his Institute in Culture and Creation Spirituality of teaching suspect theology, it is also true that one finds there a willingness to speak for the whole created order in ways that it was not spoken about at St. Anselm or in theology in general.

Out of my experiences in two very different institutions, I arrived at a position I call the New Catholic Mysticism. It calls us to faithfulness to integral Catholic truth, while bringing the moral power of that truth to bear on the crisis that threatens God's green earth. We can be faithful and active in the promotion of sound ecology while being Catholic Christians. In fact, under present conditions, to be faithful I believe we must be ecologically alive, or our Christian faithfulness is suspect.

In the creed we say that "we believe in God, the Father almighty, Creator of heaven and *earth*." In the liturgy we use Ambrose's hymn and say, "All *creation* rightly gives you praise, all *life* and holiness comes from you." At the offertory of the Mass, we say, "Through your goodness we have this bread to offer, fruit of *earth* and the work of human hands." Why then are we continuously forced to deal with hostile camps, one on the right denying the sacrality of the created order, while they praise the Creator, and another on the left, ridiculing, ignoring, trivializing, embarrassed about the thrust of three thousand years of God's revelation in the Jewish and Christian traditions? Can we not have both spirituality and earthiness?

Those who encourage us to be drunk with self-centered pieties while ignoring the destruction of God's green earth sin against earth, against animals, and against their own children. They lead us on to collective suicide and blaspheme the very God they praise, the God who found creation good, who so loved the world that he sent his only Son. Those who ridicule and trivialize the concern of Jewish and Catholic traditions in upholding the sanctity of life *in utero* replace truth with egocentric railing against the God of covenantal love and blaspheme against the work of the Spirit in time. They usher us on from the promised land of revealed love to a desert of meaninglessness, and in doing so they encourage us toward individual despair and suicide.

For twenty centuries we've processed to the table of God to eat. At their core Roman Catholic, Orthodox, and Anglo-Catholic churches hold out the incredible claim that we can eat God! They hold that Jesus as Christ is co-eternally God with the Father and the Spirit, that he is the Logos of the universe, of all cosmologies new or old, the author of the galaxies, earth, life, and cultures. Last of all, he appeared in the story, was put to death, rose to new life, and left a covenantal meal that would provide his very self for us to eat.

If the above is not true, then Christianity is the greatest hoax, the most foolish caper, the most marvelous absurdity the world has ever seen. But if true, then no other reality measures up to it.

The church, though, suffers from the same collective bipolar disorder that threatens all of Western culture. Still, however polarized the church may appear—or even *is*—what about the alternatives? Fundamentalism? New Age movements? Charismatic renewal? Pre-Vatican II Catholicism? Arrogant masculine authoritarianism cannot replace genuine religious experience. Exaggerated feminist rhetoric doesn't help much either. The gay community's alternative isn't viable for heterosexuals, who remain the vast majority of humankind.

As the arrogant and the superficial polarize, they tear the great curtain in two. Through that curtain shines a *third option*—the Triune God shining forth in the eucharist, the axis of the cosmos and the unwobbling pivot of world Christianity.

Will we heal the earth without the author of the earth? Will we heal the life community without the support of the author of life? I hear the commotion from right and left and a host of ideologies claiming that each—radical feminists, restorationists, extreme pro-abortionists and extreme pro-lifers—enjoys privileged status as representing the most important agenda. What really happens is that we live on a see-saw of alternating claims to represent the universal way forward. Jesus the Christ, though, is the one Christians know as the author and Logos of the cosmos, beyond the dichotomy of right and left, a true *third option* beyond parties.

When the planet appears to be dying, when animals disappear, when preborn children are disenfranchised and subject to genocide, when life becomes burdensome and violent, perhaps we should consider our options. Admit that we just don't know how to live. That we can't put things back together. That we've lost the instructions and can't remember the plot. Perhaps we need to consult the author? Perhaps we need to surrender to the author and discover the paradoxical truth that in surrender we find freedom as God's sons and daughters.

In the beginning, before the great cosmic fireball, was the Word. *Embracing Earth* is an attempt to put between two covers the insights and wisdom of persons who are part of a New Catholic Mysticism that believes Jesus is that Word, God's very *sophia-*

hokma-wisdom incarnate, the carrier of the great cosmic code that is ultimately saving mystery, good news for earth and all who dwell in her.

Notes

1. Donella Meadows, Dennis Meadows, and Jorgen Randers, *Beyond the Limits: Confronting Global Collapse, Envisioning a Sustainable Future* (Post Mills, Vt.: Chelsea Green Publishing Company, 1992).

2. Eugene Odum, *Fundamentals of Ecology*, 3d ed. (Philadelphia: Saunders, 1971).

3. Barry Commoner, *The Closing Circle: Nature, Man and Technology* (New York: Alfred A. Knopf, 1971).

1

Earthly Offerings

Sacrifice and the Land in Ancient Israel and Early Christianity

FREDERICK G. LEVINE

Three times a year you shall hold a festival for Me: You shall observe the Feast of Unleavened Bread—eating unleavened bread for seven days as I have commanded you—at the set time in the month of Abib, for in it you went forth from Egypt; and none shall appear before Me empty-handed; and the Feast of the Harvest, of the first fruits of your work, of what you sow in the field; and the Feast of Ingathering at the end of the year, when you gather in the results of your work from the field. . . .

The choice first fruits of your soil you shall bring to the house of the Lord your God. (Ex 23:14-16,19)

In these twilight days of the twentieth century, when the earth seems on the verge of giving its last gasp under the weight of the ecological onslaught that humans have wrought upon it, the Western religious traditions—Judaism and Christianity—have been made favorite targets of the environmental battle cry. Ever since the opening salvo by Lynn White, almost thirty years ago,[1] these two religions, but especially Christianity, have been singled out as the fount from which the philosophical and indeed theological rationalization for the rape of the planet has sprung.

And yet, how can this be so?

How can the religion of ancient Israel, which for a thousand years revolved around the ritual offering of crops and animals, be accused of having no connection to the land? Similarly, how can Judaism's daughter religion, Christianity, which was founded upon the idea that God took on flesh and blood to walk the planet, be considered otherworldly?

The fact is, if we look closer and get beyond the over-simplified distortions of secular critics and the hyper-spirit-ualized interpretations of religious moralists, we will see that Judaism and Christianity, in their own ways, have put as high a price upon the planet as any other spiritual system. Contrary to the prevailing critique, it is not Western religion that is to be blamed for the current alienation from the planet, but rather society's *rejection* of religion in the name of capitalism and industrialization, accompanied by a gross misunderstanding of the symbol systems of the Western religious traditions.

Pagan Sacrifice in the Near East

In order to get a clearer picture of the role of the earth in Western religion, we need to go back to the roots of the tradition and focus on the practices of the people who first considered themselves Jews; for it is the fields and vineyards of ancient Israel that served as the birthplace for what eventually came to be present-day Judaism and Catholicism. And it was through the sacrificial cult of ancient Israelite religion that the relationship to the earth was most clearly expressed.

But first, we need to examine the backdrop against which the Israelite nation came into being. We know from archaeologists, as well as from the Hebrew scriptures, that sacrifice was a common form of worship in the ancient Near East. However, the motivation for that sacrifice was quite different from that which underlay the Israelite practice. What we see when we look closer is a worldview that was motivated by fear. Sacrifice in this context often took the form of an attempt at appeasing the forces of nature that held sway over agricultural peoples.[2]

One function of sacrifice in the ancient Near East was the feeding of gods and goddesses—incinerating animal carcasses so that they could be made "edible" for the deity. Another type of sacrifice was aimed at appeasing the higher powers—offering up valuable property in exchange for divine beneficence. The desperation that drove such practices can be seen in the fact that humans, frequently children, were sometimes sacrificed in addition to animals. In such cases, the sacrificer would be making a bargain with the god: "Take this life in place of mine, and ensure that the coming year will be kind to me and the rest of my kin (tribe/city/nation)."

Finally, at its most basic level, sacrifice functioned to ensure a good harvest in the coming year—rain at the appropriate time and no pestilence upon the crops. In a world that depended upon a bountiful harvest for its survival (not just the survival of individuals, but the survival of the religious, political, and military institutions upon which society rested), it was essential to take all appropriate steps to try to influence the fate that nature had in store. Thus, the first fruits were often given over to the deity, as a way of acknowledging that the best portions of the harvest (and the herds) rightfully belonged to the higher powers as payment for continued human survival.

So, although the practice of sacrifice has deep roots throughout the Near East, as well as throughout the ancient world, the motivations for such practices were often based on fear. Rather than a worldview that saw humanity and nature in perfect harmony, living in a garden of earthly delights, the stark reality was a world filled with hostile and fickle forces, where humans saw themselves at the mercy of powers that were at best unconcerned with human welfare, and at worst violent and malevolent.

The Land in Ancient Israel

Against this backdrop came the ancient Israelites, who, although they also based their religious life on the practice of sacrifice, saw that sacrifice in a different context. What made this people different is that they postulated a single God, not just a God who was stronger than all the rest, not just a God who worked his punishment and benevolence through the auspices of nature (as in the parting of the Red Sea), but a God who *transcended* nature, who transcended the material world—to the point where the people who worshiped him were not even allowed to contemplate a graphic image of the deity. This God was not *of* the world, but *above* it, in a way that broke radically with the previous understandings of divinity. And yet, oddly enough, this God, too, required earthly sacrifices from his human followers.

Critics of Judaism and Christianity are quick to seize upon this transcendent aspect of God as proof that Judaism (and later Christianity) was built upon the notion of an unbridgeable gulf between God and nature.[3] God exists wholly apart from nature, the argument goes, and since, according to the Bible, humans were created in the image of God, humans, too, are seen as disconnected from their natural environment. In this view, creation functions as a mere backdrop for the divine drama that plays out between God and humanity.

While it is true that Judaism's primary contribution to the Western world was its radical monotheism, such a theological thread is far from the whole story. To begin with, we have to return to the context into which this idea was introduced. As we have seen, for the Near Eastern pagan world, nature was something to be feared. The Israelites, like their neighbors, were an agrarian society—their lives revolved around the cycles of nature and the hopes and fears that made up such a fragile existence. It is ludicrous to think that by emphasizing the absolute transcendence of God, as the Hebrew scriptures do, the ancient Israelites would have convinced themselves that they were above nature—that they were no longer dependent upon the ebb and flow of the seasons in their proper time.

To those of us living in the modern industrial world, severed from the absolute dependency upon nature that characterized human existence up until a few hundred years ago, it is perhaps too easy to get a distorted view of the Bible's message. Just as modern environmentalists struggle against the tide of our current disregard for nature, so, too, did the ancient Hebrews face a monumental task in their effort to counteract the prevailing notions of their time—that nature was a divine, often malevolent entity who had to be worshiped and appeased, frequently at human expense.

For the ancient Israelites the land was not only the source of their very existence, but it represented the cornerstone of their relationship with God as well. The first thing their transcendent God did in establishing a special relationship with Israel was to promise them land.[4] The land became the foundation of the covenant between God and Abraham, and through Abraham, with the rest of Israel down through the ages. This awesome and mighty God, who brought Israel out of slavery and swatted the Egyptian war machine like a fly, does not promise his chosen people eternal life, a place in heaven, or even dominion over the rest of the world. Rather, their birthright as God's holy people is a thin slice of land bordered by the Jordan River and the Mediterranean Sea.

For Israel's part, the people's half of the covenant is to swear unending obedience to God, which they do at the foot of Mt. Sinai, where they witness a mass revelation, the nature of which is unmatched in world religious literature.[5] The product of that meeting between Israel and God is the Torah, which, according to Jewish belief, was given to Moses at the same time that he received the Ten Commandments. Although the Torah technically refers to the Pentateuch, it is more accurately defined as the detailed set of moral and ritual prescriptions that make up the entire Jewish corpus. The covenant is clearly stated: If Israel abides by this code of behavior, her reward will be the land that was promised to Abraham. Moreover, the people will be "fruitful" and go on to fill the land with their abundance. These are very earthly rewards.

The laws of the Torah can be divided roughly into two categories—those dealing with the way in which people relate one to another, and those governing the relationship between God and Israel. Here we see an intertwining of both the horizontal and the vertical realms of human existence—morality and spirituality. Thus it is not just religious life, but also the way in which society should ideally function, that is mediated through God. As we shall see, the sacrificial religion that grows out of the Torah serves to intertwine the moral and the spiritual at the same time that it grounds both dimensions of human existence firmly in the earth.

Probably the most pronounced example of the sacred character of the earth in the Hebrew Bible is to be found in the laws concerning the Sabbatical and Jubilee years. According to ancient Jewish law, every seventh year is a Sabbatical year, during which the land is to lie fallow. This is true, as well, of the Jubilee year, which occurs every fifty years, and at which time all land seized as payment for debts is to be returned to its original owner. The message is clear: the earth is the Lord's, not the property of individuals, not an asset to be picked up by a bank, not something to be foreclosed upon. Moreover, the bounty of the earth is also the result of God's beneficence, and, to remind themselves of that fact, every seven years Israel is commanded to leave the land as they found it. Additionally, when reaping the harvest, each landowner is obligated to leave a corner of the field uncut, so that the poor and disenfranchised can come and glean what is left.

These laws do more than keep feelings of greed and possessiveness in check; they also serve to tie in the responsibility that each member of society has for every other member. All participate equally in the covenant and, by virtue of the covenant, in the promise of the land. Thus, on the human-to-human level, the land itself becomes a means by which justice, fairness, and righteousness are played out between people. These are not abstract notions or ideals that belong only to a perfect world or to some kind of messianic future. Rather, the Torah has found a way to embody these ethical ideals in the reality of the earth itself.

Serving God through Sacrifice

But it is when we come to the laws governing the relationship of humanity to God that the role of the land is absolutely central. As we mentioned earlier, that relationship was based upon obedience, and obedience most prominently took the form of ritual sacrifice. The intricate and detailed laws governing the times and methods of sacrifice take up much of the space in the Pentateuch; as such, they relate directly to the Temple in Jerusalem. But even earlier, in the book of Genesis, we see examples of the way the earth served to sanctify the relationship between God and Israel.

Specifically, we read of frequent encounters between God and the patriarchs—Abraham, Isaac, and Jacob. Sometimes those encounters took the form of "dialogues," sometimes the form of visions, voices, or dreams. In many cases, the result was that the recipient of such revelations would erect a stone or ring of stones or some form of altar from the rocks of that particular location in order to immortalize that place as a sign that God had been experienced there. Thus, right from the beginning, pieces of the earth were considered sacred enough to represent a divine encounter.

The most famous such story is the binding of Isaac, in which Abraham is commanded by God to sacrifice his "favorite son."[6] At the last minute, when Abraham has bound Isaac upon the altar and is about to slit his throat, an angel's hand reaches down from heaven and stays the blow. Abraham looks up and sees a ram caught in a thicket, which he then sacrifices in Isaac's place. Here, God has supplied his own sacrifice, and it is in a very earthly form. Moreover, the location of that sacrifice, Mr. Moriah, later becomes associated with Mt. Zion, the Jerusalem Temple mount. (We will have more to say about the connection between Isaac's near-sacrifice and the sacrifice of Jesus below.)

The institution of sacrifice reaches its zenith with the establishment of the Temple in Jerusalem. Now, instead of independent altars set up throughout the countryside, the regular cycle of festivals is observed in a central location. This is more than just a geographic arrangement. The Temple, where all sacrifices are to take place, is constructed on the top of Mt. Zion, which supersedes Mt. Sinai as the locus of interaction between God and humanity. Whereas Mt. Sinai represented the point where the covenant was renewed and codified in the form of the Torah, Mt. Zion is the place where the covenant is to be fulfilled. Israel has inherited the land. It now remains for the people to keep their half of the bargain and worship the Lord according to the detailed instructions set down in the law.

Mt. Sinai is but a way station on the journey to the promised land; the Temple is the permanent home for God's presence.[7] As God's dwelling place, Mt. Zion also represents the cosmic mountain, the *axis mundi*, the intersection of heaven and earth. It is the place where God and humanity meet.[8] But despite the majestic proportions and awesome grandeur of the Temple, it nevertheless contained very earthly elements. The entire interior, we read in the Bible, was lined with the best cedar imported from Lebanon. Moreover, we are told that no metal tool was allowed to strike the stones that made up the external structure while setting them in place. Similarly, the altar stone itself, where public sacrifices took place, had to be an unhewn stone—which was seen as perhaps a piece of the cosmic mountain itself.

Much has been made of the fact that the Temple represented a microcosm of the earth.[9] The pillars at the entrance may have stood for the pillars that were thought to hold up the firmament of the world, separating the "waters above" from the "waters below." In the courtyard there was a huge basin of water supported by twelve carved oxen, symbolizing the sea. The doors were of carved olive wood, as were the cherubim that adorned the holy of holies. And the entire edifice was finished off with carved flowers, pomegranates, palms, gourds, lions, and oxen.[10] This, then, was the stage for the regular meeting of God and Israel.

The sacrificial calendar, set out in painstaking detail in the Pentateuch, made up the cycle of festivals around which the religious life of ancient Israel revolved. That calendar was based upon the agricultural calendar; it began in the spring and ended in the fall. The major feast days consisted of the three so-called pilgrimage festivals—Passover, Shavuot, and Sukkot—during which people from all over the land would bring their sacrificial offerings to the Temple in Jerusalem. It is not coincidence that these feast days corresponded to key harvest times in the seasonal cycle of the Near East.

Passover coincided with the Feast of Unleavened Bread, which came in the spring and marked the barley harvest. Shavuot, or Feast of Weeks, which came fifty days later (and thus is also known as Pentecost), commemorated the wheat harvest. And Sukkot, or Feast of Booths, was celebrated in the fall, at the time of the harvesting of tree fruits. All these holidays were marked by the offering up to God of the first fruits of the harvest—barley, wheat, olives, wine, or oil—but they also consisted of animal sacrifices, often the firstborn of the flocks and herds, which were slain, burned, and eaten, except in some cases, when the entire carcass was consumed by fire.[11]

A key element of these sacrifices was the blood. The Israelites were commanded in no uncertain terms that they were not to consume the blood of animals, because blood is life. The sacrificial process included slitting the animal's throat so that the blood would run out into vessels. While the carcass was then burned or cooked over a fire, the blood was poured into receptacles that ran down into the ground.[12]

From a mythological point of view, we could speculate that the life of the animal was being returned to the earth whence it came. Thus, a form of recycling of the life-force was part and parcel of the sacrificial process. On the one hand, the animal was offered up to God, the giver of all life; at the same time, the earth, as the engine that fuels the life cycle of the planet, was symbolically "refueled" with the blood.

However, there was another aspect to the disposing of the blood. In some types of sacrifice, the blood was dashed upon the altar as a kind of ritual cleansing agent. What was being washed clean was the sins of the priests, who were performing the sacrifice, and the sins of the Israelite nation as a whole. The Bible saw sin as almost physically polluting, a force that had to be neutralized before God could be present to receive the sacrifice.[13] Thus, what we see here again is an intertwining of the moral and the spiritual.[14] That is to say, sin—the moral life of the people—had a *physical* effect on the Temple, an effect that could only be eliminated through the use of blood—the life force of the planet. Thus, the earth itself, through the offering up of animal life, acted as a sanctifying agent for human moral transgression.

Sukkot, in particular, marked the end of the agricultural year and, to this day, includes prayers for rain in the coming winter months in order to ensure a bountiful harvest. In addition, Jews were commanded to construct "booths," or tabernacles, in which they were to dwell for seven days. These huts were temporary shelters unable to withstand the onslaught of nature. They still make up an integral part of the observance of Sukkot, and today they are decorated with the harvest bounty. In place of a weatherproof roof, they are covered with green branches through which the stars are visible. Thus, the fragility of life in the face of the natural world, in addition to the earth's fecundity, is a central theme of this holiday.

A key point to bear in mind is that, despite these strong agricultural associations, the Torah did not see the performance of ritual sacrifice as a way of ensuring a bountiful harvest and good weather, as the pagans did. To be sure, there must have been a magical element to the practice, which was not lost on the ancient Israelites. But from the point of view of the Torah, nature was not to be manipulated by human beings, since God was the absolute ruler over all of creation. (That is not to say that the Torah did not accept the efficacy of magical practices—in fact, they are specifically outlawed.)

The careful performance of sacrifice was Israel's way of fulfilling its side of the covenant. Through it, the people showed obedience to God's commandments. In return, it was not so much a good harvest that resulted (although that surely made up a part of God's promise), but rather God's love and protection, and, of course, the continued possession of the land. When the Hebrews were exiled from their land, beginning in the sixth century B.C.E., they saw their banishment as a form of punishment for not keeping the commandments.

To underline the fact that these were not simply agricultural festivals, the Bible gave each one a twist: Each of the agricultural feast days was associated with an historic event that marked a milestone in the history of the Israelites. Thus, Passover recalled the Exodus from Egypt, Shavuot eventually came to commemorate the giving of the Torah at Mt. Sinai, and Sukkot was seen as a reenactment of the way in which Israel lived in impermanent huts or tents during their forty years of wandering in the wilderness.[15]

The message here was that it is not nature that determines the fate of humankind, but God. By acting through history—some would say, by intervening in history—God not only exercises absolute sovereignty over creation, but also demonstrates love for humanity. By recalling the Exodus from Egypt, the giving of the Torah, and the forty years of wandering in the wilderness, the Israelites were acknowledging the intricate relationship that exists among God, creation, and humanity.

The remarkable achievement of ancient Judaism was that it found a way to posit a God who was absolute, omnipotent, and transcendent, and yet establish a relationship of love between such a God and humanity. What made that relationship possible—what validated the covenant—was the institution of sacrifice, which in turn was fueled by the fecundity of the land. In other words, it was the land itself that established the means of ongoing communication between God and Israel. It is as though pieces of God's own creation were "recycled" back to God, the source from which they originally sprang, through the act of ritual sacrifice. In the process the land itself was sanctified. The earth was thus doubly holy: first, because it is God's creation, but secondly, because it continues to play an integral role in the love song between God and humanity.

Thus, from the biblical point of view, creation was not merely the backdrop to the divine-human drama, as some critics of biblical religion have charged, but rather the vehicle by which the human-divine relationship was made possible. Moreover, rather than nature being subordinate to history, from this point of view it is history, through the mechanism of the covenant between God and Israel, that now acts as the means by which creation is redeemed. Instead of the simple reductionist model that has often been imposed upon the Bible, what we have, in fact, is an exceedingly intricate web, made up of God, nature, humanity, and history, which functions as a closed, self-generating loop, and from which no single part can be separated out.

The biblical attempt to combine a historical or time-developmental overview of the world with the cyclical aspect of nature seems especially relevant given our current understanding of the nature of the universe—that the world began with an initial moment of creation (the Big Bang) and continued to develop over time through the process of evolution (from the evolution of galaxies, stars, and planets to the evolution of plants, animals, and humans).

If evolution is seen as moving forward, then the ancient pagan view of nature as simply going around in cycles is incomplete. Ancient Judaism saw history as moving forward in a meaningful way, and although the understanding of creation was not sophisticated enough to articulate a theory of evolution, in its own way this attempt to superimpose historical meaning upon what was perceived as a directionless natural world can almost be seen as a rudimentary attempt to incorporate nature into the purposeful flow of history, an unconscious attempt, if you will, at an evolutionary teleology.

The Sacrifice of Christ

Following the destruction of the Jerusalem Temple by the Romans in 70 C.E., the sacrificial cult came to an end. As a result, the rabbis faced the daunting task of revisioning Judaism without the Temple as its center point. They did this by focusing on prayer, charity, and repentance, and were aided by the prophetic tradition, which had always stressed the importance of morality over empty ritual action. When the first-century rabbi Johanan ben Zakkai (who is credited with rebuilding Judaism after the fall of Jerusalem) sought to reassure a colleague who had burst into tears at the sight of the ruined Temple, he told him that the performance of sacrifice was unnecessary as long as they could perform deeds of loving kindness.

The heroic effort of the rabbis to salvage Judaism from the ashes of Mt. Zion was paralleled by that of nascent Christianity, which was struggling at the same time to piece together a purely spiritual religion that nevertheless shared a continuity with the Jewish biblical tradition. Ironically, the notion of sacrifice came to play a much more central role in Christianity than it did in rabbinic Judaism. (Perhaps because traditional Judaism still looks forward to the reestablishment of the Temple, and therefore the sacrificial cult, it did not seek to "replace" the performance of sacrifice with specific deeds.)

The notion of sacrifice manifested itself most clearly in two elements of early Christianity: the sacrifice of Jesus Christ and the eucharist. Both of these aspects of Christian thought exhibit strong links with the way in which sacrifice functioned in ancient Israelite religion.[16] In fact, there is evidence that the earliest Christian communities still practiced the sacrifice of first fruits. Instead of being offered up to God, however, they were given to the prophets, who took the place of the Temple priests. If no prophets were present, they were given to the poor. Thus, we can see the same connection between ethical norms and ritual sacrifice that we saw in biblical Judaism. This trend was developed even further, however, by drawing upon the strong prophetic moralizing tradition in an effort to spiritualize the notion of sacrifice. What is significant for our purposes is that the idea of *material* sacrifice was not given up entirely by the early church. The land, therefore, continued to play a significant role in the relationship between God and humanity.

But the idea of sacrifice in Christianity reached its apotheosis with the death of Jesus. Whereas Judaism had managed to historicize the meaning of sacrifice through the pilgrimage festivals, Christianity went even further and eschatologized it through the person of Jesus Christ. It is no coincidence that Jesus is referred to as the "first fruits," and Paul's Letter to the Corinthians clearly draws upon a nature metaphor in pointing out that in order for resurrection to take place, the "seeds" must first die. That is, the fruits of the harvest must rot in order to take root once again. Thus, in the same way that the products of the land (which come from God the Creator) were recycled back to God through sacrifice, so, too, is Jesus (who comes from God) recycled back to the Father through the sacrificial process.

What is different now, though, is that Jesus represents the perfect sacrifice. Rather than a sacrifice on behalf of an individual, which must be repeated on a cyclical basis, the death of Jesus represents the sacrifice of all humanity once and for all. Thus, each person participates directly in the great loop of creation-death-resurrection through the material body of Christ. The key point here is that even in a religion such as Christianity, which sought to spiritualize the notion of sacrifice, in order for salvation to take place—in order for the cosmic loop to be closed—*God had to take material form.* In the same way that the land played an indispensable role in the sacrificial loop of the Temple cult, so, too, do the "first fruits" of the material world afford the very mechanism by which the world is sanctified in Christianity.

But Jesus is not only referred to as the "first fruits." He is also called God's "favorite son," the term being a direct reference to Isaac, Abraham's "favorite son." Here we see a complicated symbol set that involves the sacrifice of Isaac, the death of Jesus, and the role of the Paschal lamb as a primal sacrifice cementing the relationship between God and Israel.[17]

In the Book of Exodus, on the eve of the departure from Egypt, the Israelites sacrifice a lamb and smear its blood on their doorposts so that the angel of death, who on that night slays the first-born of the Egyptians, will pass over their houses.[18] Clearly, the lamb represents a kind of substitution for the firstborn sons of the Jews. Similarly, earlier, when Isaac is spared at the last minute, Abraham sacrifices a ram in his place.[19] The place of that sacrifice, Mt. Moriah, comes to be associated with Mt. Zion, where the Temple is later built and the sacrificial cult institutionalized. Thus, a homologation takes place between Mt. Moriah, Mt. Sinai (the destination of the Jews upon leaving Egypt), and Mt. Zion, which involves the sacrifice of the firstborn and the Paschal lamb.

In the Gospel of John it is clear, due to the timing of the Last Supper, that the crucifixion takes the place of the Paschal sacrifice, and Jesus becomes the "Lamb of God."[20] (It is significant that Jesus dies before the Romans have the opportunity to break his legs, since any sacrificial offering must be perfect and entirely free from blemishes.) Thus, for John (and for Paul as well), Jesus is introduced into the Isaac-Paschal lamb-firstborn son matrix. Jesus thus represents the continuation of the biblical sacrificial tradition.

This means that the covenant, which as we saw was so central to the sacrificial role of the land, is now continued through Christ. Or, in Christian terms, the old covenant made at Sinai (and sanctified through the Paschal lamb offering) is now replaced by a new covenant made at Golgotha (and sanctified through the sacrifice of Jesus). Both the Passover offering and the death of Jesus represent acts of supreme obedience to God, which, as we have also seen, are the means by which Israel was to keep its half of the covenant with God. The blessing of Abraham, which initially passed on to Isaac and through him to all Israel, now passes on to Jesus and through him to all Christians. That blessing, it should be remembered, centered around the promised land.

Eucharist: The Sacrifice of the World

This intricate symbolism all comes together in the eucharist, the symbol of Christian sacrifice par excellence. Initially, the eucharist was considered a first-fruits offering, especially because it quite literally consisted of the fruits of the harvest, grain and wine. But because of its central role in the Last Supper, it also came to be associated with the Paschal lamb offering, and consequently with Christ himself.

For Paul, this meant that the eucharist was a way of entering into communion with Christ in the same way a communion sacrifice in the Temple cult was a way of entering into relationship with

God.[21] The material bounty of the earth therefore becomes the vehicle by which humans can draw closer to God. But for Christianity this also meant that through the eucharist, the individual could participate directly in the sacrifice of Christ. The degree to which the eucharist was seen as a sacrifice of the type practiced in the Temple is reinforced by the fact that initially one had to be in a state of ritual purity in order to partake of it, in the same way the priests officiating at the Temple (as well as those bringing sacrifices) had to be ritually pure.[22]

Finally, the eucharist moved beyond the idea of communion, or participation in the sacrifice of Jesus, and came to represent an actual theophagy, the eating of Christ's body.[23] Again, this act of eating is part of the sacrificial typology of the Temple cult, in which most sacrifices were apportioned between the sacrificer and the priests and eaten. In the case of Christianity, however, because Christ is both spirit (as God the Father) and matter (as Jesus the son), the notion of eating the sacrifice gets mixed in with the notion of eating God.

Similarly, Christ is both the sacrificer (the Father offering up his "favorite son") and the sacrifice (the son). As a result, by participating in the eucharist, the Christian becomes both the sacrificer and that which is sacrificed. Here, the idea of redemption takes a new turn. For, whereas in Israelite religion the fruits of the land represented the means by which humans participated in redeeming the land, in Christianity people themselves are drawn up into the redemptive process as part of the sacrifice. Again, in both cases, the material world becomes the means by which that redemption takes place. And although Christ's sacrifice is often seen as the perfect, and therefore last, sacrifice—the sacrifice to end all sacrifices—it is clear from the ongoing sacrament of the eucharist that the redemption of the world takes place sacrificially again and again, just as it did for the ancient Israelites at the Temple.

It is perhaps remarkable that all this complex sacrificial symbolism, bound up in ancient Judaism and Christianity, could be reduced to the simple fact of eating bread and drinking wine. And for a religion that began as an attempt to spiritualize the worshiping process, to build that process around such material symbols seems a bold and paradoxical step. We can therefore not lose sight of its extreme significance—that the land and its life-giving bounty played a central role in Christianity, just as it did in ancient Judaism.

Notes

1. Lynn White, Jr., "The Historical Roots of Our Ecologic Crisis," *Science*, vol. 155, no. 3767.

2. For an overview of general theories of sacrifice, see Henri Hubert and Marvel Mauss, *Sacrifice: Its Nature and Function* (Chicago: University of Chicago, 1964).

3. In addition to Lynn White, Jr., more recent critics include feminist writers who view the place of women in Judaism and Christianity as a parallel expression of the treatment of nature and the Deep Ecology movement, which has drawn upon the nature mysticism of the East as an antidote to Western alienation from the natural world.

4. For example, Genesis 12.

5. Exodus 19.

6. Genesis 22.

7. Jon D. Levenson, *Sinai and Zion: An Entry into the Jewish Bible* (San Francisco: Harper & Row, 1985).

8. Mircea Eliade, *The Sacred and the Profane: The Nature of Religion* (San Diego, New York, and London: Harcourt Brace Jovanovich, 1959), pp. 20-65.

9. Raphael Patai, *Man and Temple* (Ktav Publishing House, 1967).

10. See 1 Kings and 2 Chronicles.

11. For a description of the different kinds of sacrifices performed at the Temple and their function, see Baruch A. Levine, *In the Presence of the Lord* (Leiden: E.J. Brill, 1974); and Roland de Vaux, *Studies in Old Testament Sacrifice* (Cardiff: University of Wales Press, 1964).

12. Ibid.

13. Levine, p. 63.

14. This intertwining of the spiritual and the moral is seen most clearly in the description of the Day of Atonement ritual found in Leviticus 16. For a detailed description of the use of blood as a purifying agent, see Jacob Milgrom, *The JPS Torah Commentary: Numbers* (Philadelphia: The Jewish Publication Society, 1990).

15. See Irving Greenberg, *The Jewish Way: Living the Holidays* (New York: Summit Books, 1988), pp. 15-118.

16. The following analysis is based primarily upon the work of Frances M. Young, *The Use of Sacrificial Ideas in Greek Christian Writers from the New Testament to John Chrysostom* (Cambridge, Mass.: The Philadelphia Patristic Foundation, Ltd., 1979).

17. For a detailed treatment of this relationship, see Jon D. Levenson, *The Death and Resurrection of the Beloved Son: The Transformation of Child Sacrifice in Judaism and Christianity* (New Haven and London: Yale University Press, 1993), pp. 200-232.

18. Exodus 12.

19. Genesis 22.

20. John 1:29; 19:31-37.

21. Young, pp. 240-56.

22. Ibid.

23. Ibid.

2

God, the Cosmos, and Culture

ALBERT J. LaCHANCE

The space of the Universe was in the shape of a hen's egg. Within the egg was a great mass called no thing. Inside no thing was something not yet born. It was not yet developed, and it was called Phan Ku.

Nana Buluku, the Great Mother, created the world. She had twins, Mawa and Lisa. She did nothing after that.

In the beginning, God was Wul-bar-i. And God Wulbari was heaven—spread not five feet above the mother, earth.[1]

Cultures blossom forth from sacred stories. What God did in fashioning the universe, earth, the life community, and culture, we repeat ritually in order to fashion our human world. Humans and human cultures seem to have this one great fact in common: we intuit that our origins are sacred. Sacred stories create a context for human life. They provide a niche in that they articulate right relationships with the plants, the animals, and the living systems that surround us. The stories explain the origins of the night sky. The heroes and villains of the stories are mirrored back to us by the constellations. Terror, order, beauty, and chaos are all explained by the story. Our stories surround us in the institutions that blossom from them. They are psychic nests. When these nests are destroyed, as was the case with the plains cultures of North America, the culture enters chaos, panic, hysteria, cultural pathology. The only hope in such a situation is to allow the nests to

15

be recreated. To recreate such a nest, we must return to a life lived in imitation of God. We must live again within a cosmogonic story.

From time to time cultures grow weak in their own self-understanding, and they experience the equivalent of a collective identity crisis. At such times foreign or diseased stories can enter a culture and function as psychic pathogens. Examples are apartheid, Communism, black slavery, and Native American genocide. In referring to the singularly grotesque example of Fascism in Germany, C. G. Jung coined the term *psychic epidemic*. He was comparing the spread of insane ideas through a culture lacking the immune system of a viable story to the spread of infectious diseases such as the bubonic plague. It might be helpful to think of the condition as cultural AIDS. Fascism ended in the same way that psychic epidemics always end, with death and destruction of both the human and natural worlds. It ended as horribly for Japan. But following the devastation and death came resurrection and new life.

Presently we find ourselves in what Thomas Berry calls a deep cultural pathology. We need only look into what T. S. Eliot referred to as "the strained time-ridden faces, distracted from distraction by distraction," in order to see and feel the deep trouble lurking in the Occidental soul. Our understanding deepens when we consider the rubble that characterizes vast areas of so many of our cities. But in our day the problem goes deeper. Complete populations of plants, insects, and animals disappear daily. The atmosphere of the planet is fraying. Soil systems and water systems are dying. The context for renewal, the nest itself, is unravelling. Because human cultures are themselves nested in these systems, human cultures are being undermined. Thus our collective feelings of hysteria, stress, and despair.

Why do we foul our own nest? Because we are not functioning within our own origin story. Secularism, not itself a story, but rather the absence of a story, has invaded the cultural mind. Secularism is the denial of the values that characterize our true story. With this disintegration of values, and the devaluation of life itself and the context for life, come a whole host of demon ideas such as the "right" to exploit life forms to extinction, the "right" to exploit living systems to their death, the "right" to limitless sexual expression and the "right" to destroy the living beings thus conceived. These bogus rights exist only within the moral vacuum of cultures emptied of sacred story.

As the stressors placed upon the natural systems become more virulent, there are corresponding stressors placed upon human individuals. With toxic environments come intoxicated people. With species destruction come dizzying levels of ever more horrendous violent crimes. With habitat destruction comes alienation from the natural world, a loss of context, terror. Epidemic levels of addiction, hatred, racism, warfare within the habitat of the human womb, despair, and suicide are just some of the human symptoms associated with planetary collapse. While hysterical advertising supports our denial of our real condition, the psychiatric and stress reduction industries profit from our breakdown. If human life is nested within the larger life systems of the planet, then how will human stresses decrease when the stressors placed upon the living systems increase? How will human psyches be healed while the psyche of the earth disintegrates? All the "have a nice day" smiles, all the exhortations to "keep it positive," all the frenzied attempts to feel good or to pretend to feel good, all the "I'm okay—you're okay," all our starry-eyed "praise the Lords" can only mask our symptoms in the short term.

The principal religions associated with the West are Judaism and Christianity. Most of us intuit that somehow their fire has dimmed. Castrated of the prophetic function, both seem uncomfortable with the affective presence of Yahweh, in particular with the divine outrage. In both, clerical functionaries scurry to appear relevant by adopting questionable techniques from the behavioral sciences. Meanwhile, the voice of the Creator is silenced, creation is plundered of its life forms, the vitality of the whole life community dims, and our own zest for life evaporates. "Religious authorities" refer seekers to psychiatry for mood elevators rather than lead them on the journey to wholeness that they themselves have not taken. Young people who hear the Sacred Voice calling them from within are unable, without direction, to name it as such. Some are even diagnosed as delusional. They hunger for spiritual and moral authority; we offer them condoms and "political correctness." They know that the earth is dying; we "protect" them from that interior truth. They want to grieve the death of the animals; we tell them that they are too sensitive, that they can't change the world. The result of all of this? Despair. Addiction. Suicide.

Those young who have survived their parents' "choice," look to their parents, ministers, and teachers for some concern for life. Too few of us have been willing to give our energies to the task of creating the future. Too few of us have been willing to support the young in building the foundations in their characters that will sustain them in the questionable future they face. Having lost the model of the Divine Parent, many of us seem to be losing the ability to parent properly. The loss of authentic, spiritually grounded moral and ethical values has rent the souls of Occidental people. There's a great hollowness in the hearts of the young. They feel this hollowness acutely but find difficulty in naming it. They need us to validate the agony they feel. But we find it difficult to accept responsibility for the state in which we find ourselves. It is easier to blame our traditions for having let us down. Perhaps there is some truth in this. But, as T. S. Eliot prophetically asked in *Choruses from The Rock*,

> You, have you built well, have you forgotten
> the cornerstone?
> Talking of right relations of men, but not of
> relations of men to GOD.
> "Our citizenship is in Heaven"; Yes, but that
> is the model and type for your
> citizenship upon earth.

Can we blame our traditions? Can we blame God or our traditional understanding of God? We know from all Twelve-Step programs that any sincere appeal to God is a sufficient starting place for recovery from addiction. Would God be any less responsive were we to call out for healing of our culture and its relationship to the planet? Only our refusal to cry out prevents receipt of a response. With the loss of trust in God and in our traditions comes the loss of faith, energy, and the zest for life. With our cheapened view of covenantal religion, Jewish and Christian, we've lost functional communication with the Divine Parent. With that loss goes our sense of tradition, the link that binds the present to the past and the past to the future. We've lost the familial continuity that can only come from blood lines. We're orphaned! Small wonder our children find belief in the future difficult. They feel lost, orphaned by heaven and earth. Perhaps we're hitting bottom. Perhaps we are crying out.

Because there are huge issues to face, many of which we have mentioned here, it might seem like the worst of times. But in many ways, it is the very best of times. Thomas Berry's work has provided us with a connective understanding of history. In what he calls the New Story we've all been given a history which includes the history of the cosmos, earth life, and culture as a single, splendid multiform sequential emanation. As such, the New Story has four chapters. In the first chapter we read of the birth and evolution of the cosmos from the fireball to the formation of the galaxies and our solar system. From the solar system emerges the Earth Story, including plate tectonics and the creation of the atmosphere, the oceans, and land masses. From the Earth Story comes the story of life from prokaryotic to eukaryotic, from fish to birds, from reptiles to mammals, from primate to human. And emanating from this Life Story comes the multiform story of human cultures, tribal, neolithic, classical, modern, and postmodern.

Everything in the cosmos, living and nonliving, participates in the cosmos chapter of the story. The whole earth, living and non-living, participates not only in the cosmos story but shares uniquely as well in the Earth Story. While all living and nonliving beings of earth share in the Life Story generally, the particulars of the story begin to differentiate at the continental, bioregional, and ecosystemic levels. The Life Story of Africa is different from that of Australia, while they both participate in overall biogenesis as a whole. In the cultural phase of the story, differentiation becomes highly nuanced. Still, at the interior level of culture there seems to be the hope of transcultural experiential identity. That interior ex-perience is activated by means of a cosmogony. Cosmogonies sacralize the universe and provide symbols through which that sacrality can be experienced.

Cosmogonies are stories that tell us how the universe came to be. They are the psycho/spiritual ground from which cultures spring. They give us our understanding of divine transcendence, and they provide access to the immanent sacrality of the universe, the earth, life, and culture process. They tell us how God fash-ioned creation. Mimicking God, we create a human world that provides us with a niche within the natural world. Moral action comes to be understood as action that mimics the actions of God; immoral action is action that contradicts the action of God. Truth is seen to be self-evident, because it refers to the story of how God acted. Not only our religions but our cosmology, education, econo-mies, politics, and healing systems spring from our stories of God, our cosmogonies. Without the ongoing ritual reenactment of the cosmogony, the story, the organizing seed of the culture, with-ers; the cosmology along with the culture that bloomed from it decays. Naturally, the individual disintegrates along with his or her culture. Occidental culture is presently experiencing such a decay.

While the new cosmology provides a superordinate basket within which all local cosmologies can be nested, each of the local cul-tural stories must nonetheless be wed to it if the sacrality of the cosmos/earth/life/cosmos process is to be accessed. What good is it to know that the cosmos is a sacred process if we cannot experi-ence and thus access that sacrality as a power for transformation? In the Occident, our cosmological self-understanding has bloomed from the cosmogonic seed of the Parent-God Yahweh. Our history turns on the cosmogonic axis of a biblical cosmogony. In the re-cent past our understanding of ourselves has become progressively confused. Certain elements within our cosmogony and thus our cosmology have come to be seen as the root cause of widespread ecosystemic damage associated with the industrial process. But even as this awareness dawned on us, the corrective was already being revealed. Out of the Occidental seedbed had sprouted the new cosmology. Since the new cosmology springs from the old Occidental cosmogony, they are both ultimately biblical at root—phases one and two of the same cosmology. Both find their source in and are the expression of the Parent-God Yahweh and the Human-God Jesus Christ.

The entire Occidental cultural enterprise, including its most recent scientific phase, is grounded in the Yahweh/Jesus cosmogony. Yahweh was and is the transcendent, pre-fireball stage of the cosmogony, and Jesus Christ as Logos of God is its immanent phase. The Spirit is the communicator throughout creation, allowing the persons of Creator and creature to be known to each other. This communication is covenant, and the breach of covenant is the severing of communication. The severing of communication from within the creation creates the condition of orphanism. The Parent-God is present and speaking, but we can no longer hear. This is our present state of affairs. This is the source of our emptiness.

In losing this relational love called covenant, we've lost functional relationship with the Parent-God, with the universe, the earth, life, and with each other. The glue has loosened; we live in a dying light, the twilight of love. Timothy Ferris is fond of quoting Einstein's statement that "the most incomprehensible thing about the universe is that it is comprehensible." With the loss of covenant and communicability the cosmos becomes incomprehensible, opaque. The place of splendid beauty becomes a place of terror. Family relationships that are no longer built on covenantal love are replaced by breakdown and alienation. Individual and cultural pathology is the result of this loss. The disintegration of our spiritual, mental, emotional, physical, and ecological health results from this loss. Einstein is also said to have wondered how God thinks. Perhaps his ability to listen to the thoughts of God created for him the comprehensibility of the universe. Our further disintegration into widespread madness is guaranteed if we do not return to our cosmogony and the two-stage cosmology that has bloomed from it.

Does the Western tradition have the interior resources to inspire and support a response to our present crisis? I submit to you that the New Story wedded to our spiritual cosmogony is that response! The New Story emerges from the deep interior of the Occidental soul. It is the explicit expression of the infinite potential of God boiling up from within the Occidental tradition. Our scientific understanding of the fireball is the empirical confirmation of the light in Genesis and in John 1. As Rupert Sheldrake observes:

The broad outlines of the Genesis myth and the contemporary scientific account are not dissimilar; they have a strong family resemblance. The scientific account is of course far more detailed, and attributes creativity to chance rather than to God. But both, by their very nature as accounts of origins, refer to events that happened before there were people to witness them and can therefore only be imagined, calculated, inferred, or modelled. They can never be statements of observable or observed facts.

The creation theories of science have grown up within the Hebrew-Christian cultural matrix, with its paradigmatic conception of a beginning, a Fall, historical progress toward the end of history, and an end that in some sense reestablishes the beginning. The theory of the Big Bang and the modern doctrine of universal evolution bear a striking resemblance to this fundamental myth of our culture. (It may not be a mere coincidence that since we now live in fear of ending our civilization in a cataclysmic nuclear war, we have come to a creation story that begins with a vast explosion.)[2]

Timothy Ferris, in his P.B.S. film series *The Creation of the Universe*, similarly points to the relationship between our idea of *uni*-verse and *mono*-theism. Even the science of ecology is prefigured in the biblical story of Noah's ark.

Noah: The First Deep Ecologist

In Genesis 6:6-8 we read: "Yahweh regretted having made human beings on earth and was grieved at heart. And Yahweh said, 'I shall rid the surface of the earth of the human beings whom I created.'" But it's not until we come to verse 13 that we discover the true cause for God's disgust and regret. There God says to Noah: "I have decided that the end has come for all living things, for the earth is full of lawlessness because of human beings. So I am now about to destroy them and the earth."

Is this not the exact situation in which we find ourselves today? Is not the whole living community of earth threatened because of violence of human making? How often are we warned by the data fed back to us by the living systems of the earth, as when rising global temperatures threaten to melt the polar caps causing the oceans to rise and to flood the dry land?

Happily for Noah, and for the life community he protected, he had a special relationship with the person of God. Covenant is incomprehensible if God is not a person. Covenant with Tao,

Dharma, Nam Myoho Renge Kyo, or with the implicit order is confusing. A person can relate to any of these, but without a relationship with the person behind the idea, it remains essentially opaque. Only if God is personal, not an old man with a long beard beyond the clouds but a precosmic and intracosmic thinking, feeling, and responding Spirit, can covenant make sense. But in that case, covenant is everything one could hope for, dream of, the reply to every desire, an intimate relationship of love with the living God of love who fashioned the universe and dwells, radically present, within it. Moreover, if Jewish law can be thought of as cognitive, left brain and masculine, then covenant is relational, connective, right brain, and feminine. If the decalogue expresses divine kingship, then covenant expresses the queenly aspect of heaven. So we read in Genesis 6:18-22:

"But with you I shall establish my covenant and you will go aboard the ark, yourself, your sons, your wife, and your sons' wives along with you. From all living creatures, from all living things, you must take two of each kind aboard the ark, to save their lives with yours; they must be a male and a female. of every species of bird, of every kind of animal and of every kind of creature that creeps along the ground, two must go with you so that their lives may be saved. For your part, provide yourself with eatables of all kinds, and lay in a store of them, to serve as food for yourself and them." Noah did this; exactly as God commanded him, he did.

In Genesis 7:2, a further concern for all creatures is expressed:

"Of every clean animal you must take seven pairs, a male and its female; of the unclean animals you must take one pair, a male and its female (and of the birds of heaven, seven pairs, a male and its female)."

In Genesis 7:14-16,

With them [was] every species of wild animal, every species of cattle, every species of creeping things that creep along the ground, every species of bird, everything that flies, everything with wings. One pair of all that was alive and had the breath of life boarded the ark with Noah, and those that went aboard were a male and female of all that was alive, as God had commanded him.

The ark, then, is the ecojudaic symbol par excellence. It is a metaphor for the whole earth and for Yahweh's love of the whole earth. Again, the ark is the relational, the covenantal, the womb of the Divine Parent. Noah's ark is what we now call Gaia. And God's concern doesn't end with the preservation of the life community. Following the flood, Noah is enjoined to be concerned with the freedom of the animals to propagate, with their right to habitat. This right is a truly divine right, mandated by the Divine Parent. In Genesis 8:15-19 we sense this maternal tenderness as the animals are ushered out of the ark:

> Then God said to Noah, "Come out of the ark, you, your wife, your sons, and your sons' wives with you. Bring out all the animals with you, all living things, the birds, the cattle and all the creeping things that creep along the ground, for them to swarm on earth, for them to breed and multiply on earth." So Noah came out with his sons, his wife, and his sons' wives. And all the wild animals, all the cattle, all the birds and all the creeping things that creep along the ground, came out of the ark, one species after another.

Happily for the animals and plants, verse 21 contains a promise of Divine protection: "Never again will I curse the earth because of human beings, because their heart contrives evil from their infancy. Never again will I strike down every living thing as I have done."

The natural world is afforded covenant with God and protection by God against humanity. This bodes ill for us presently, because it represents the reestablishment of right relationship between Creator and creatures, between cosmogony and cosmology, a covenant our death-dealing modernity seems hell-bent on breaking. In Genesis 9:6 we read:

> He who sheds the blood of man,
> by man shall his blood be shed,
> for in the image of God
> Was man created.

That verse should send a shudder through those who would kill "for God" as well as those advocating and/or performing abortions. In both cases, human blood is spilt by choice. The Noahic Covenant establishes that all life, that all species, that all life systems are sacred. The Creator reasserts the original cosmogony reactivating the sacrality of creation, of the whole creation. Life is sacred or it is not; we cannot have it both ways. To be against abortion and indifferent to ecology is schizophrenic. To be an ecological activist and support the present genocidal attitude toward human life in its womb is equally mad. God will not have it both ways. Life is sacred.

In Genesis 9:8-17 a divine encouragement is given to all living beings, human and nonhuman. Relational love is established between the Creator and all creatures. Humans are told to mimic the attitude and actions of God. We are to be in covenant with the earth, affording her and all her creatures the same concern and protection afforded them by God. Because we've abandoned our own relational love of God, we've likewise abandoned our relational love of the animals and plants.

> God spoke as follows to Noah and his sons, "I am now establishing my covenant with you and with your descendents to come, and with every living creature that was with you: birds, cattle and every wild animal with you; everything that came out of the ark, every living thing on earth. And I shall maintain my covenant with you: that never again shall all living things be destroyed by the waters of a flood, nor shall there ever again be a flood to devastate the earth."
>
> "And this," God said, "is the sign of the covenant which I now make between myself and you and every living creature with you for all the ages to come: I now set my bow in the clouds and it will be a sign of the covenant between me and the earth. When I gather the clouds over the earth and the bow appears in the clouds, I shall recall the covenant between myself and you and every living creature, in a word all living things, and never again will the waters become a flood to destroy all living things. When the bow is in the clouds, I shall see it and call to mind the eternal covenant between God and every living creature on earth, that is, all living things."
>
> "That," God told Noah, "is the sign of the covenant I have established between myself and all living things on earth."

Thus is Noah anointed by God as ecology's patron saint. Perhaps those of us still professing biblical faith should begin to pray to Noah on behalf of the whole created order. The example of Noah provides a model for Noahic Jews and a Noahic order of Christian brothers and sisters of creation. Mother Teresa once said that if you want to call young people to service, ask them to give all of themselves. The young hunger to give themselves to something real. Occidental religious life will blossom again only when the call made to the young is an eco-spiritual call, a call to save the plants, insects, animals, and habitat of the earth in the name of their Creator. The stale rags of formalism, ecclesial, corporate or educational will no longer allure them. The young, like God, want life.

If the Noahic story is ever to be activated, if we are ever to understand ourselves, we must accept our unique history, our unique theology, our science, our story. If we are to have an experience of the cosmic-earth-life-culture process as a whole, we cannot do so cognitively only. To decouple the New Story from the cosmogony of its source is to imagine another desacralized universe. To tell the Occidental story without the Yahweh/Logos cosmogony is to tell not our story, but an abstraction, a cognitive construct. Occidental spiritual experience springs from a personal and parental revelation of the sacred. Cosmology without cosmogony is scientific gnosticism.

When the cosmos-earth-life-culture process is experienced as emanation from God, each element of the process is experienced as sacred, sacramental. History is the story of God's Spirit creating and interpenetrating the cosmos, earth, life and all human cultures in a progressive revelation. There are no so-called secular cultures. The secularizing of culture is the cultural pathology we've considered here. All culture is sacred because in the beginning was the Word. A refusal to integrate our whole story as it has been revealed in religion and in science will have the same result as when addicts refuse to deal with reality. Such unfortunates are condemned to a slow and degrading death precisely because they are unable to deal with their real experience, their real story, their lives as they really are and really have been. Our refusal to integrate our whole history condemns us to remain locked within the frame of reference of our present predicament. Our pathology can and will only worsen as it becomes our future.

To remain functionally separated from our own story, whether in its cosmogonic or its cosmological phases, is to remain stunted, out of center, addicted, orphaned. Many addicts and many "religious people" who are in fact religion addicts, know a great deal about the spiritual life, but can only become sober and well following a spiritual experience. Most who do get well refer to a moment of truth when the grace of a power greater than themselves entered their lives. Twelve-Step programs such as the one offered by Alcoholics Anonymous provide a structure through which truth and grace can broaden and deepen from spiritual knowledge into spiritual experience. The absence of such spiritual experience is the fundamental cause of our cultural and planetary plague. Experience springs from cosmogony; experience sacralizes cosmology. From sacred cosmology spring sacred geology and biology.

Now, while it seems clear that Occidental culture springs from an understanding of God as parent, it is equally clear that too great an emphasis has been placed upon the masculine attributes of the Divine Parent. Perhaps this was necessary while the masculine revelation differentiated itself from the much earlier feminine revelation. In any case, if the present hyperpatriation cannot be blamed for all our present woes, it has certainly been the cause of some and has exacerbated others. If the Occident is to move toward renewed evolution, toward renaissance, we must encourage the *rematriation* process. While a renewed emphasis upon native traditions and Jewish and Eastern Christian mysticism will help, a heavy burden of creativity falls upon us all.

Masculine ecclesial solutions will not help much here for obvious reasons. Politically motivated solutions such as the democratization of ordination might not be the surest road either. There is, after all, no such thing as a "female priest." There are priestesses with their own unique power for ministry. To overlay women with the persona of a traditionally masculine role might temporarily satisfy our hunger for political fairness, but there is a limited good in this. In the longer arc, it seems to be just another masculine answer to a problem of hyperpatriation. Instead, a new Hebrew and Christian priestess must be born from the seedbed of the Occidental tradition.

The creation of ideational constructs that attempt to change Occidental religious structures by extrinsic overlay, by manipulation, or by superficial political notions of fairness seem doomed to failure. The feminine must be understood cosmogonically, cosmologically, geologically, biologically, and culturally, and then supported in a renewed evolution. While the political approach to change in religious matters seems superficial, the attempts to resuscitate past revelations of the sacred seem equally superficial. With all due respect to Robert Bly, "Zeus energy" isn't a religious reality with which many of us identify. Propping up anachronistic cosmogonies can seldom serve anything but tangential purposes.

Equally unhelpful is the blandness of spirit that results when religions lose their grounding in the created order. Willfully blinding ourselves to the horrors of ecological devastation while lulling congregations to sleep with starry-eyed pieties is sinful. Avoiding prophetic confrontation with economic empire is cowardice. Praising the Creator while we ruin creation is blasphemy. What is needed is a spontaneous resurgence of our love of life itself. We must all fall in love again, with each other, with each other's cultures, with our born and preborn children, with all the other life forms, with earth, the cosmos, and God.

Perhaps the greatest hope in this regard lies among the vast numbers of "baby boomers" who have journeyed outside the Hebrew and Christian traditions. Many of us have spent years and even decades studying Eastern and native religious traditions. Were all these explorers to return with their treasures to Western religions we might be able to see ourselves and our traditions from a completely unique vantage point. This new ingredient could be just what we need to reenergize ourselves, perhaps even sparking a renaissance. Failing in this regard, the sad legacy left by my generation might be epidemic addiction and sexually transmitted diseases, legally sanctioned mass crimes against preborn human young, faithlessness, alienation, consumerism that dwarfs pre-sixties levels by whole orders of magnitude, and a much vaunted political correctness that feels ever more like Fascism to those disenfranchised from its premises. Through mass media the culture has been inundated with moral toxins until the very fabric of Occidental culture has frayed. We have idealized decadence and created mass media, rock 'n' roll anti-heroes.

Regaining the Mandate of Heaven— Our Way toward the Good Life

Confucionists have a teaching called the Mandate of Heaven. It refers to a condition wherein the governed of Chinese society truly believe that those governing are doing so from a position of moral truth. Even when times are difficult, if the governed believe that their difficulties are not caused by the moral turpitude of the powerful, they will grant their support to the government. That support is the mandate of heaven. Conversely, should the people become convinced that this is not the case, the government will lose their support and fall. This collapse might be forestalled by violence, but becomes all the more inevitable should violence be applied. We just watched astonished while the Soviet government lost the mandate of heaven. Soon, the government in the land of Confucius will change or will experience a similar fate. It might be easy to congratulate ourselves were we not in a similar situation, not only within Occidental societies, but with the governance of the natural world itself. In our cultural pathology we are losing the mandate of heaven and of earth.

Can we possibly feel secure while our institutions crumble around us? While the natural systems globally languish and groan? While the air unravels? While the soils disintegrate? While the oceans, rivers, and lakes die? While millions die needlessly in the name of our supposed right to limitless sexual expression without regard to our personal or collective responsibility? While the first president from my generation opens his administration by legalizing experimentation on the bodies of the victims of that bogus right? While we eat, breathe, and drink violence, and bring our children to birth in it, and then nurse them with more? While plundering economies further devastate the planet? While we mutilate our souls by denigrating our powerful and beautiful traditions? While we mutilate the bodies and souls of others in other traditions? While we go on refusing to return stolen lands to the native peoples? While we invent ever more lies to believe in? Can we expect the Mandate of Heaven while we deny heaven and plunder the earth? When is enough enough?

For more than a decade Thomas Berry has called for a cultural therapy. I'd like to outline some ideas that I think must be central to any attempt to realign the culture with adaptive practices. First, we must be willing to shatter all forms of denial that defend us from experiencing our denied fear and disgust concerning our present eco-cultural predicament. Second, we must be open to hearing the voice of the sacred calling to us from within our own cultural traditions. Third, we must surrender utterly to God. The whole history of the Jews could be summed up in the pulsations between being lost to Yahweh and found through surrender to Yahweh. Fourth, we must make a complete list of our crimes against life, crimes against the soil, water, air, and life forms, human and nonhuman, born and preborn. Fifth, we must confess these lists to a trusted confidant in order to shed the unconscious plague of repressed guilt. Sixth, we must make a list of the mental and emotional habits that continue to motivate us to so much violence. Seventh, we must ask God to remove these habits.

Eighth, we must make a list of all persons, groups, species, and natural systems we are harming. Ninth, we must make it our personal mission to contribute to the healing of life. Tenth, we must continue to critique ourselves and each other with respect to our bio-friendliness or the lack of it. Eleventh, we must fully reenter the journey of faith, making moral health our primary life goal. Twelfth, having reexperienced the Mandate of Heaven as a result of these steps, we must become willing to live a good life and to exhibit leadership in the healing of the earth. How shall we define a good life? I would say that a good life, a life well lived, is one that enhances the vitality and variety of living forms that succeed it, one that passes on a viable and healthy culture to those who inherit our culture and the planet. Therefore, a good life is one that is surrendered to God, to God's will on earth, among other species, among cultures, and among one another. There doesn't seem to be any other way; there never was.

As all this takes place we shall experience a renewal of Occidental culture. Wedding our cosmogony to our new cosmology we shall all become storytellers, and our story will go something like this:

Before the beginning was God. When God willed light, the fireball appeared. God's Word was within the fireball, and that Word became matter. From matter the galaxies came forth. From the galaxies the earth appeared. From earth came water, and in the water the Word became living cells. From living cells organisms appeared—fish, birds, reptiles, insects, mammals, flowers and fruit, marsupials and primates. The Word became human and told stories about God. From out of the stories grew tribal, neolithic, classical, and technocentric cultures. The Word revealed science, and we walked upon the moon. The Word became love and insight and meaning in many places and in our time today. Today, the Spirit of God is upon us, inspiring us toward a new understanding of wholeness, of God, of the earth, and of culture.

Notes

1. These epigraphs are taken from Virginia Hamilton, *In the Beginning: Creation Stories from Around the World* (New York: Harcourt Brace Jovanovich, 1988), pp. 21, 43, 53.

2. Rupert Sheldrake, *The Presence of the Past* (New York: Times Books, 1988), p. 258.

3

Christ the Ecologist

JOHN E. CARROLL

"All of us are children in a vast kindergarten," Sebastian tells her, "trying to spell God's name with the wrong set of alphabet blocks."
—Tennessee Williams, *Suddenly Last Summer*

If we consider something that might be called the Christian ethic, not necessarily as practiced by the majority of those calling themselves Christian or who are perceived by others as Christian, but as put forth by the initiator of Christianity and his early disciples some two millennia ago, the central Christian tenets or ethic are overwhelmingly ecological, overwhelmingly in synchrony with the principles of ecology. There appears to be no discrepancy. Before considering those tenets or precepts, however, it is worthwhile to keep two thoughts in mind.

First, when that great philosophical ecologist and student of Christianity Mahatma Gandhi was asked what he thought of Christianity, he replied that he thought it would be a good idea. (He is reputed to have said the same about Western civilization.) The implication is, of course, that Christianity does not exist in practice—only as a concept on paper. If it did exist, the world would be a very different place.

Second, it has been said that Francis of Assisi is the only Christian who ever lived, at least among figures publicly known. (Christ, of course, was a Jew, and the word *Christian* means "follower of Christ.") Once again, a mere glance at the central tenets readily reenforces that notion. Not coincidentally, Francis also would have to be viewed as perhaps the purest of ecologists or ecological thinkers.

30

The central precepts of Christianity are at least fourfold. And they are all fundamentally ecological, indeed exceedingly so.

- love your neighbor as yourself
- avoid worship of false idols
- avoid the sin of pride (or false pride, as it is often expressed)
- live simply

Students of religion will recognize that these ideas are not limited to Christianity—they appear in many other religions, other faiths—but they are indeed central to any and all definitions of Christianity.

Love neighbor. But who is our neighbor? Most people today take this to mean fellow human beings. There was a time not too long ago (and continuing in some places today) when *neighbor* excluded indigenous or native peoples; certain people of color; non-European peoples in general; and, to some extent, women and children. Many humans today are still somehow lesser beings to some of us—they are not quite our neighbor in an equal sense. Technically, legally, we generally accept the notion (in theory but not always in practice) that human beings have intrinsic value. Hence, human ownership of other human beings is unacceptable. (Yet the most brutal forms of exploitation and repression can well be found acceptable, forms worse than some forms of slavery.)

Let us look also at the other side of the coin. On what grounds can we put forth the premise that *neighbor* should be limited to other members of our species? How can we be sure, especially in light of our scientific and ecological knowledge, that other animals, plants, indeed the inorganic universe of water, air, and rocks are not our neighbor? We are totally dependent on all these things, and modern physics and ecological science both teach us that we are one in relationship with them. So, how can they not be our neighbor? Can we be so sure of our narrow interpretation of this term?

We are commanded to love our neighbor and our God. Where is this God of ours? Is it going too far to think that God is in Creation? Perhaps we should express the word *creation* with an upper-case "C." Does not God reside throughout? Is it so farfetched to think of God as immanent in all Creation? I think not. There is much support in Christian scriptural interpretation, Protestant and Catholic, to suggest "God in all." Must we not, therefore, love Creation if we are to love God? Hence, a broadening of *neighbor* and a recognition of God in Creation seem necessary. The impact of truly following this command would be profound. Or should we avoid this profound change and stop calling ourselves Christian?

Avoid false idols. This is the very first of the Ten Commandments, and the one we perhaps gloss over or ignore more than any other. This precept conjures a Hollywood image of ignorant, primitive people worshiping the sun, ritualistically burning fires, or sacrificing animals before large sculptures, perhaps even taking human lives in such sacrificial manner. These pictures are so nonsensical in modern times that we conveniently come to think that this precept no longer is relevant. Yet the problem, a profound one, perches right in front of us, and we choose not to see it. It is money as a bottom line, as an end in itself. Money is the false idol, and we do indeed pay it homage.

Either a society believes money to be the bottom line, the end rather than a means to an end, or it doesn't. It has been said that any society that forgets money as a means and uses it as an end has signed its death warrant; it will achieve its demise in short order. Sadly, all appearances indicate that money is, for us, for our society, an end in itself, the bottom line. We have collectively as well as perhaps individually broken the first commandment. We worship the idol of money, a false idol by any reasonable interpretation of scripture. Money takes the place of God, Creation, and neighbor.

The impact of dislodging the god Money from on high, of replacing it with God in our neighbor, with God in Creation, would be profound, ecologically, socially, and spiritually. Or should we avoid such profound change and stop calling ourselves Christian?

Avoid false pride. False pride is also called overweening pride, arrogance, hubris, or vainglory. Some Christians are so concerned about this, including the Amish of America, that they see pride as the central problem underlying all others. They see this tenet, therefore, as the central one. The Amish, among some others, see the sin of pride not only as overbearing and universally present in all, but beyond our control to deal with as individuals. Indeed, the very notion that we could individually overcome the existence of sinful pride is itself a sin of pride and compounds the existence of the original sin. The Amish see the overcoming of the sin of pride only through community, through peer pressure, through people helping people, by their presence exerting a controlling influence to keep pride in check. As Christians they have a basic need to be "Christians in community," to acknowledge this basic precept of Christianity, and to do something about gaining control over it.

Over the years the level of Christian concern over the sin of pride has fallen and the word itself gained weakening qualifiers such as *overweening, false,* and *excess.* I don't believe the Amish would use such qualifier words. To them and to earlier Christians the sin is pride, not overweening pride or false pride or excess pride. In our drive to put ourselves on a pedestal as superior to all, we have been much encouraged by the Cartesian-Newtonian-Baconian thinking and the mindset of the European Age of Enlightenment.

This mindset says that we are central and dominant, that we are superior, that we know best, and that we should use our power to mold, manipulate, command, and control nature, Creation, and ultimately God. The world has become our toy, there for us to tinker with and play around with, to use as we would. We have become separate from Creation, and somehow above it. We yielded to the temptation of the sin of pride, the temptation to power, and have savaged the planetary ecosystem, many fellow human beings, and generations yet unborn.

A reversal of this attitude, an acceptance of what ecology tells us—that we are part of something larger which gives us our lives and asks us to be responsible in return—would lead to a very different world, a very different society, one so different as to be unrecognizable. Should we so reverse, or should we stop calling ourselves Christians?

Live simply. "Lots (of people) are discovering seeds of simplicity in their weed-patch of prosperity," observes John Daniels of Detroit. In these words we may recognize a hopeful sign of ecological restoration and recovery. Whether it comes too late or is too little is another question whose answer will become apparent at a later date.

Christianity commands—as does Buddhism, Hinduism, Jainism, and other Eastern as well as Western and indigenous peoples' thought—that we live simply, not take more than we need, and differentiate needs from wants. A popular bumper sticker today reads, "Live simply, so that others might simply live." Yes, it has come to that. So many of us live anything but simply, so much so that there is a dramatic shortage of even the essentials of life for so many others. We in the United States constitute 5.6 percent of the world's population but demand and take well over 25 percent of the world's energy and natural resources. While there is no lack of others, including Europeans, who take more than their share, we North Americans (Canadians as well as ourselves), are far and away the biggest demanders, the biggest takers, on a national and on a per capita basis. Even our relatively poor population takes quite a bit by world standards.

The global population question is often at the forefront in any environmental or natural resource conservation discussion, but in the United States the focus of such discussion almost always revolves around numbers of people (usually far-away third-world people) and not our own per capita consumption. It is obvious that we want to both scapegoat other people and focus on seemingly insolvable problems, so that we don't have to *do* anything. We want to avoid focus on ourselves, on our own behavior, lest we have to do something hard in light of such focus. We quickly steer the discussion, in other words, away from living simply.

We also readily confuse living simply with asceticism, and in that way dispose of it as unreasonable and unacceptable. Asceticism is certainly one way of living simply, but that is not the argument. Indeed, it is a deliberately gross distortion of the argument, designed to avoid the issue. Given that we seem to have what some call an addiction at the societal level, an addiction to consume and consume and consume, the precept to live simply cannot come easily. And if the notion of living simply comes into vogue as an idea, our response undoubtedly will be more consumption: we would buy books that tell us how to live simply, audio tapes for our car, and video tapes for our home, books and tapes to instruct us as if we couldn't otherwise figure out how to live simply. Eventually, we would toss out our expensive and complicated goods, but only to replace them (causing even more consumption) with simpler products, at least until the fad passes.

The question we fail to address is, What constitutes a high quality of life? We all want a good life. We all strive for it. But how do we define it? It can just as readily be defined as a life with less, as it can as a life with more. And, generally speaking, the fewer possessions, the simpler the life and the more freedom. The more we have, the more we need to worry about, to protect, to maintain, and, increasingly today, to worry about disposing of. Why can we not challenge ourselves to see how little we can make do with? How about seeing if we can mathematically reduce our demands, our wants, our needs, each month? Those who win are the ones who can make do with the least, who can simplify the most! On the surface this appears to be sacrifice, making do with less. A little probing underneath, a little experience, however, suggests it is anything but a sacrifice. New horizons of joy and pleasure readily can open before the person who successfully gets rid of the clutter. Of course, the society overall experiences a major change in this reversal of behavior, in this reversed value system. And, if widespread through society, the old society collapses and a new one rises in its wake. The one which rises is infinitely more sustainable. It is ecological, healthy for all God's creatures in a way that the present society can never be, and it is Christian in the highest sense. Can we truly simplify, can we separate needs from wants, can we support such a societal change, or should we stop calling ourselves Christians?

The behavior of that portion of the world's humanity which calls itself Christian appears to violate all of these central precepts of Jesus Christ. Was Gandhi right? Would Christianity be a good idea? Would the world be a better place if there were a few Christians around?

4

Open to Life—and to Death

The Church on Population Issues

DAVID S. TOOLAN, S.J.

When Pope John Paul II declared in his World Peace Day message of January 1, 1990, that "education in ecological responsibility is urgent" and insisted that the "right to a safe environment" should be included "in an updated Charter of Human Rights," the statement was no doubt welcomed by the Sierra Club and the Worldwatch Institute. The pope is no James Watt. Yet under their breath, I suspect few environmentalists were convinced that the pope had thrown the weight of the Catholic church behind their cause. Something was conspicuously missing: Not once had the pope mentioned the link between the world's exploding population growth and environmental degradation. And though few demographers and ecologists now claim that birth and fertility rates are the only—or even the principal—factor to be concerned about, they certainly deem population stabilization a key element in any constructive program of environmental conservation. That the pope avoided this issue seemed to many observers to be of a piece with the Holy See's pattern of challenging or minimizing the significance of population trends at UN population and development conferences at Budapest (1974), Mexico City (1984), and Cairo (1994), not to mention the Earth Summit in Rio de Janeiro in 1992.

Background

Shortsighted economic development, environmental degradation, and irresponsible fertility behavior, most experts agree, have a three-way, reinforcing interrelationship. No one factor can be taken in

35

isolation. This was the underlying assumption, at any rate, of the three-year-long study by the UN's World Commission on Environment and Development chaired by Prime Minister Gro Harlem Brundtland of Norway (also known as the Brundtland Commission, published as *Our Common Future* in 1987), which concluded that the human future is at risk if we continue the current unsustainable modes of economic growth and development, and if these are replicated in developing countries.[1] The Brundtland report led to the Rio Earth Summit and its new watchword: *sustainable development*. The term means the generation of wealth and the use of resources within the limits of ecological possibility. Good ecology, it was argued, is good business. In effect, governments were being called upon literally to change the bases of economic life (or to "ecologize the economy"). The Earth Summit action plan, spelled out in a list of concrete measures called Agenda 21, assumed a set of new priorities for the world community, among them:

- To revive the economies of developing countries, which means reversing their net outflow of resources and ensuring their access to the global commons, additional resources, and energy-efficient technologies.
- To eradicate poverty, the principal source of environmental problems of developing countries and a major threat to global environmental security.
- To find patterns of production and consumption in the industrialized world that will reduce their disproportionate contribution to the deterioration of the earth's environment.
- To meet the essential needs of the developing world's expanding population, while working to stabilize world population.
- To reverse the destruction of renewable resources, soil, forests, and biological and genetic resources through alternative means of production and consumption.
- To devise new accounting procedures for gross domestic product so that it measures resource depletion, and to change the system of incentives and penalties to encourage sustainable economic development.
- To merge environmental and economic concerns in decision-making at all levels.

Generally, the Catholic church fully supported this set of priorities. Certainly it had far fewer qualms than President George Bush did when he heard that the annual cost of Agenda 21 programs might reach $125 billion. The Vatican delegation only became nervous when efforts to stabilize population were mentioned. But just because of reservations on this point, many of the participants in this discussion—particularly Western diplomats, feminists, and environmentalists—look upon the church, along with Senator Jesse Helms, as an obstructionist.

Is There a Demographic Problem?

The Vatican plays down the role of overpopulation in ecological destruction. At best, it concedes it is a problem in "some" areas. And where this problem exists, church representatives typically say the remedy is not a crash program of family limitation, but an attack on poverty and the unjust social conditions that force the populous poor to devour the land. The church has a point, but most development experts I talk to do not think the population issue can be so easily set aside. It is simply too big—and has to be faced straightforwardly.

I imagine that most readers of this book are familiar with the argument of environmental apocalypticists like Paul and Anne Ehrlich that overpopulation will, by the next century, overload the planet's carrying capacity. A brief summary of the global demographics goes something like this: Though the human species emerged some 150,000 years ago, the alarming fact is that most of its growth in numbers has occurred in the short span of the last forty years. It took us millennia to reach the first billion in about 1800; over a century to add a second billion somewhere between 1918 and 1927; about thirty-three years to attain the third billion in 1960; only fourteen years for the fourth billion in 1974; and thirteen years to the fifth in 1987. What made the difference, of course, was the rapid decline, since World War II, of death rates, particularly in the Third World. The point is that even though birth rates in developing countries have fallen by a third since 1950, and are expected to fall by another third by 2025, world population continues to increase, adding the equivalent of another Mexico or Germany every year (that is, about ninety million). That's because the high fertility rates of Africa, South Asia, and Latin America—where nearly half the populations consists of children whose child-bearing age still lies ahead of them—have a long-lived momentum. So, even if fertility levels declined overnight to the replacement level current in industrialized countries, the population of such regions will continue to grow for another half century by 50 to 100 percent. By the year 2050, the UN Population Fund forecasts, world population will rise to ten billion people before leveling off at 11.6 billion after the year 2150. Naturally, the imponderable ravages of war, famine, and pestilence (for example, the AIDS pandemic, especially in sub-Saharan Africa) could reduce these projections. But please note the terrible cost in suffering.[2]

The question is, with 11.6 billion people competing for the planet's soil, water, and air, will the earth stand the strain— without suffering political, social, and ecological meltdown?

The basic positions on this question were first outlined two centuries ago. In 1803 an Anglican clergyman, the Rev. Thomas Robert Malthus, spurred an international debate—and invented the "dismal science" of economics—by enunciating the principle that "the power of population is indefinitely greater than the power of earth to produce sustenance for man." The classic response came from the Marquis de Condorcet, who argued that technology—"new instruments, machines and looms"—would prove capable of keeping pace with an expanding number of mouths to feed. (Thanks to nineteenth-century industrialization, Condorcet had the better of the argument then.) Today, the debate continues, with environmentalists and their biologist allies taking the dismal Malthusian line; they tend to adopt something like Paul Ehrlich's I=PAT formula, meaning that environmental impact equals population times affluence (consumption per person) times technology (damage per unit of consumption). If we don't act quickly—and perhaps coercively—for the common good, they say, the I=PAT equation spells ecological doomsday.

Not so, cry many contemporary laissez-faire economists. Echoing Condorcet, they have faith in science and technology (and for once, in the managerial skills of governments) to keep the earth pumping to meet the demand. Almost none of the dire predictions of Mr. Ehrlich's bonanza best-seller of 1968, *The Population Bomb*, they point out, has materialized. The similar projections of the Club of Rome in *The Limits of Growth*—that the world would run out of oil in 1992 and arable land in 2000—seem equally exaggerated. Neither book figured on the "green revolution" or the discovery of Alaskan and North Sea oil in the 1970s. Bullish economists and technocrats believe necessity will be the mother of similar discoveries and inventions for the future—and the Vatican seems to join their Pollyannish company.

Who has it right—the apocalypticists or the Pollyannas? Probably neither. The correlation between population pressures and the resource base is admittedly complex and indirect. The Ehrlichs have it too simply. The environment can be degraded in locales where the fertility rate is low, such as in Eastern Europe, China, and Los Angeles, and land and water can be in decent shape in places of high population density like Hong Kong and Tokyo (if one ignores Tokyo's hand in Indonesian logging). High population density does not necessarily mean environmental damage. In the

Kakamega district of Kenya the tree density is highest where human settlements are heaviest. Conversely, low population growth and density do not necessarily lessen environmental damage. In Nepal, depopulation of the mountainsides has meant poor resource management—not enough people to maintain agricultural terraces, replant trees, and so forth.

Still, there are limits and breaking points. To accommodate an additional five billion people by the year 2050, global output, which has quadrupled since 1950, must continue to grow rapidly. But that means more energy use, emissions, and waste; more land converted to cultivation. And here one must note that traditional systems of husbandry—like letting land lie fallow for a few years to recover—can sustain people for centuries, only to collapse in a short time when population densities exceed a certain threshold. Similarly, a government capable of providing food, housing, jobs, and health care for a population growing at 1 percent a year may be overwhelmed by one expanding at a 3 percent rate (as most of Africa is). Finally, renewable resources are subject to threshold effects that belie their name. Beyond a certain point, fishing grounds that are over-fished (as in the Philippines) do not recover; extinct species (as in Brazilian rainforests) will not reappear; and topsoil (as in the sub-Sahara) is not replaceable except over geological time.[3]

The trouble is that even after the green revolution, in 69 out of 102 developing nations food production currently lags behind population growth, and about one out of five of the world's 5.4 billion people are presently so malnourished that they do not have energy for a day's work. Assuming they had work. Consider Mexico, with a population of 87 million, projected to reach 109 million by the year 2000. In rural areas, the unemployment rate often runs 65 percent; and landlessness can be near 60 percent.[4] The skewed ownership system and overworked croplands push peasants into marginal areas of highly erodible soil—and into the forested uplands. Fifty thousand square miles of the remaining forests fall victim to large-scale and small peasant farms annually. The loss of forest disrupts river flows, causing erosion that now affects two fifths of arable land, leading to the abandonment of 400 square miles of farmland a year. Desert reclaims 865 square miles of farm each year. Crop yields started declining in the mid-1970s, and in 1986 Mexico became a net importer of food. Poverty is the enemy of the environment, and in turn, nature devours her children.

The situation in the Philippines parallels Mexico's.[5] Today, the Philippines is the only Asian nation where the birth rate is rising, and the projection is a population of 115 million by the year 2020, 29 million of whom will be landless. There, too, agricultural production is falling and forests disappearing. Hardwood forestry, which in 1967 represented 32 percent of export earnings, yields only 5 percent now. And as real per capita income dropped 16 percent during the early 1980s, disaffected peasants took to insurgency movements or fled to urban slums. The Philippine bishops have come out with a remarkably fine statement on the environmental problem, but when Juan Flavier, the government's health secretary, recently launched a campaign for birth control, the bishops assured him they would sabotage it.[6]

Mexico and the Philippines are not isolated cases. Deforestation in the tropics, the most serious renewable resource decline of the past century, unravels a whole, fragile ecosystem, which effectively impoverishes over a billion people around the globe. The phenomenon is the result of a complex of factors. Population pressure is only one of them. Bad resource management—tax subsidies, pricing policy, discount rates, all in the name of economic growth—contributes heavily to the devastation. As do unjust patterns of land tenure (7 percent of the owners control 93 percent of the arable land in Latin America). What is frightening is that this havoc is endemic throughout most of Asia, sub-Saharan Africa, Central and South America. It makes one fear that we have already witnessed what the entire southern hemisphere will look like in the next century—in denuded and desolate Haiti.

If we have an environmental refugee problem now, just wait. We have to ask what the situation will be when we add another three billion people, 94 percent of them in the hungry Third World, by the year 2025? According to Dr. Nafis Sadik, the Pakistani physician who heads the UN's population agency, the numbers mean "economic and ecological catastrophe."[7] To avoid such an outcome, he calls for an accelerated campaign over the next decade in favor of smaller families. The plan would stress better education and health care for women—and could conceivably cut the mouths we have to feed in the year 2050 by 1.5 to 2 billion people. Right now, some 300 million women in the developing world do not have access to safe and reliable forms of contraception.

Environmentalists, then, are not the only ones worried. Government and nongovernmental development agencies fear that economic progress will be undermined, as it already has been in places like Rwanda and the Philippines, by high population growth. Women's rights groups come into the discussion to defend women's reproductive choice, but they also fear that a neo-Malthusian focus on reducing numbers will ignore the larger issues of women's health and economic status. Meanwhile the U.S. State Department, which during the Reagan and Bush era had cut off funding to international family planning programs, shows a new willingness to restore such financing as it translates the demographic numbers into impending political conflicts over land and water in the Middle East, Africa, and Latin America—and the prospect of more economic refugees flooding into Europe and North America.

Where Does the Church Come into This Discussion?

Very few advocates of birth control programs, I think, have much trouble with what the Catholic church promotes as an ideal of sexual morality; that sexual intercourse is not merely a recreational activity, that its proper context is a stable, loving union involving a total and reciprocal gift of one self to another; that children are a grace and parenting a high responsibility; and that, for the sake of both lovers and children, the relationship should be marked by mutual fidelity. The stumbling block lies in a few sentences in Pope Paul VI's 1968 encyclical *Humanae Vitae*. Married couples can exercise "responsible parenthood," the pope wrote, by taking advantage of the natural rhythms of fertility and infertility, but they may not use "artificial means" to hinder the procreative function. "Each and every marital act," the pope decreed, "must remain open to the transmission of life." Instead of emphasizing the total relationship or loving union as the context for judging the morality of sexual conduct, many believe that the papal teaching narrowed the focus to isolated sexual acts. Immediately, six hundred North American Catholic theologians and about an equal number of the best European theologians dissented from this teaching—and the majority of Catholics in these regions have rejected it.[8] But since then agreement with the papal prohibition of artificial birth control has become a litmus test for the appointment of bishops, and in Latin America and countries like the Philippines where the church wields plenty of political clout, politicians pay attention. As do UN diplomats.

What heads of state and UN diplomats hear from the church is well summarized in the statement of Archbishop Renato R. Martino, the head of the Holy See's delegation to the Rio Earth Summit, on June 4, 1992:

When considering the problems of environment and development one must also pay attention to the complex issue of population. The position of the Holy See regarding procreation is frequently misinterpreted. The Catholic Church does not propose procreation at any cost. It keeps insisting that the transmission of, and the caring for human life must be exercised with an utmost sense of responsibility. It restates its constant position that human life is sacred; that the aim of public policy is to enhance the welfare of families; that it is the right of spouses to decide on the size of the family and spacing of births without pressure from governments or organizations. This decision must fully respect the moral order established by God, taking into account the couple's responsibilities toward each other, the children they already have and the society to which they belong. What the Church opposes is the imposition of demographic policies and the promotion of methods for limiting births which are contrary to the objective moral order and to the liberty, dignity and conscience of the human being.[9]

In other words, the church objects when rich nations, which dominate UN agencies, dictate to poor nations, notably in the culturally sensitive area of sex and family planning. If it had its druthers, the church would prefer the UN to be neutral in this quarter—and during the Reagan-Bush years, the U.S. government adopted just such a hands-off policy.

Archbishop Martino was not fantasizing when he spoke of "the imposition of demographic policies" that violate the moral order. Indeed, many population experts now agree with him that "mistakes" were made. First of all, the World Bank has (at least in the past) resorted to a form of blackmail in this regard, tying loans to the implementation of birth-control programs. Second, even apart from the case of China's coercive sterilization and abortion program, past UN, US-AID, and private agency efforts have been deeply flawed. In India, the tactic of luring people to undergo sterilization with phony promises of money or land turned the populace against birth control. In Bangladesh and Brazil, contraceptive devices were virtually thrown at women without instruction or proper medical care. (Women often thought oral contraceptives killed sperm and were to be used only before or after intercourse.) The result was that the plan either backfired, producing greater fertility, or led to a high incidence of stroke and infection. The common difficulty is that UN-sponsored birth-control programs are often part of an International Monetary Fund austerity regime that curbs the local government's spending on education, health, and social welfare—the very things that are known to reduce birth rates.[10]

For this very reason, some feminist groups like Bella Abzug's New York-based Women's Environmental and Development Organization, have recently announced their opposition to government-imposed population control, advocating instead a "voluntary reproductive health approach with women at its center." This is not, of course, exactly what the Vatican has in mind.

Throughout the debates about internationally sponsored family planning programs in the 1970s and 1980s, the Vatican in effect threw its weight behind the loaded slogan of many third-world diplomats that "development is the best contraceptive." It meant to underscore the truism that education, health care, and employment opportunities for women are the most effective ways of lowering fertility and birth rates. (Thailand and Indonesia are shining examples that there seems to be a direct correlation between high female literacy and low fertility.)

Like many poor nations with whom it sided on this matter, the Vatican did not want a singular focus on population issues to distract attention from social conditions. For example, in a place like Haiti, ecological destruction is not just a question of numbers, but of an inequitable distribution of wealth, horrible health conditions, the abject status of women, militarism, and police repression. Indeed, high birth rates reflect a defensive reaction to grinding poverty, a kind of insurance. "The Holy See," wrote Archbishop Martino in his Earth Summit statement, "does not consider people as mere numbers, or only in economic terms. It emphatically states its concern that the poor not be singled out as if, by their very existence, they were the cause, rather than the victims, of the lack of development, and of environmental degradation."

The Vatican prefers to make affluent nations feel conscience-stricken. Third-world nations, after all, could allocate more funds to social welfare and environmental protection if they did not have to divert so much of their resources to servicing a $1 trillion external debt to Western banks. Moreover, the rich are the worst environmental offenders. With less than 25 percent of the world's people, industrialized countries consume 75 percent of the world's energy, 72 percent of all steel production, and 85 percent of all wood products; they also generate 90 percent of the world's hazardous waste, emit 74 percent of atmospheric-warming carbon-dioxide, and produce virtually 100 percent of the ozone-damaging chlorofluorocarbons. All true enough.[11]

But no one, at least in development circles, seriously questions the proposition that industrialized nations will have to change their bloated lifestyle. That's surely one major leg of the three-sided issue we are discussing here. The point is that even if the rich nations do their part in saving the planet, that won't get the Third World— or the church—off the hook of dealing with the separate issue of too many people fighting over too little land and water. Nor does it get the church off the hook of dealing with reproductive rights.

Is the Church's Stand Just?

The Vatican does not seem to have in mind, except peripherally, the often crushing burdens of unwanted pregnancies and the desperation that drive women, as a last resort, to have abortions. In the face of such problems, the official line simply calls for moral courage and self-sacrifice. Abortion is seen as a violent solution, and therefore anathema. Women are offered but one alternative, the very demanding option of natural family planning, an approach that supposes a high degree of motivation and the full cooperation of the husband. There is much to be said for NFP as furthering the equality and mutuality of spouses, and advocates of women's self-determination dismiss it too cavalierly. But whether it is realistically workable—or even just—for every couple in all circumstances may very well be doubted. Is it fair to demand such heroism, for instance, when the social and material supports for it are not in place? Or when, in the macho cultures of Latin America, the emotional support from husband and family may not be there? I think not. In such situations, it would seem morally reasonable for a woman to have available other family planning options that are safe and reliable—if nothing else, in order to avoid using abortion as a contraceptive method.

Does the church's stand on contraception, then, indirectly promote abortion? This is a very real question, as the case of Chile in the 1960s illustrates.[12] There, so many women suffered medical complications from illegal abortions that the hospitals were being flooded, leaving less and less space for live-birth mothers. The situation got so bad that doctors launched a campaign to distribute (formerly) illegal contraceptives through the national health care clinic system. In 1964 the government finally acceded—and rather quickly the crisis subsided and beds again became available for women giving birth. What this suggests, of course, is that 1) given contraceptive alternatives, women will choose contraception over abortion; and 2) the policy in Catholic countries of prohibiting contraceptive information and devices leads to an increase of abortions. Not a good move.

Somehow, church leaders do not appear to "get it"—that is, get the often desperate situation in which women, and especially poor women, find themselves—a situation where biology is destiny, where they are powerless and "have no choice." It is as if "openness to life" happens in a vacuum, apart from enabling social conditions. As one of the leading American Catholic moralists, Lisa Sowle Cahill, observes, church teaching on this point "is shadowed by the attitudes that women do not really have a right to control their fertility; that to avoid pregnancy and childbirth is to reject one's destiny of motherhood; that for women to seek roles outside of motherhood is selfish, narcissistic, and materialist, and that self-sacrifice is a specifically 'feminine' duty, especially for mothers."[13] In other words, the call for moral courage is undermined by the ambivalence of church leaders toward the international women's movement and its demand that women be recognized as integral moral agents with rightful claims to social equality and economic opportunity. Current church teaching is caught between a reluctant acknowledgment of these rights and its idealization of motherhood as pre-eminently a woman's role.

The question is how the church's ban on birth control accords with its larger social doctrine, which so staunchly asserts the obligation to be an active and productive participant in society, and society's corresponding duty to *enable* that participation. It is this social teaching which both at the top, in the person of John Paul II himself, and at the grass-roots level has made the church a potent defender of human rights and the claims of the poor throughout the world. Whether we call the poor of developing countries causes or victims of ecological destruction, the church's dilemma remains. On the one hand, a century-long tradition of church social principle exhorts Catholic bishops everywhere to put themselves firmly behind the cause of improving the education, social status, and economic position of women—and it is undeniable that nothing reduces birth rates faster than advancement in these areas. On the other hand, bishops join forces with the misogynist males of many third-world cultures in opposing the availability of those reproductive choices of which an educated and economically resourceful woman will inevitably learn. The left hand takes away what the right gives—and the policy does not cohere. Understandably, then, women get the idea that the church wants to keep them in the unequal place to which many societies relegate them.

Notes

1. *Our Common Future: Report of the World Commission on Environment and Development* (New York: Oxford University Press, 1987). On the Rio de Janeiro Earth Summit cf. *In Our Hands: Earth Summit '92/Reference Booklet* (New York: UN Conference on Environment and Development, 1992).

2. See Michael S. Teitelbaum, "The Population Threat," *Foreign Affairs* (Winter 1992/93), pp. 63-78, esp. pp. 64-66. Also Charles C. Mann, "How Many Is Too Many?" *The Atlantic Monthly* (February 1993), pp. 47-67.

3. See Jessica Tuchman Matthews, "Redefining Security," *Foreign Affairs* (Spring 1989), pp. 162-77.

4. See Norman Myers, "Environment and Security," *Foreign Policy*, 74 (Spring 1989), pp. 23-41.

5. See Myers and Matthews.

6. See "Filipino Preaching Safe Sex Stirs Church's Ire," *New York Times*, August 4, 1993, sec. A.

7. See "Curb on Population Growth Needed Urgently, U.N. Says," *New York Times*, April 29, 1993, sec. A. See also Nafis Sadik, *Safeguarding the Future* (UNFPA, 1990).

8. See Richard A. McCormick, "'Humanae Vitae' 25 Years Later," *America*, vol. 169, no. 2 (July 17, 1993), pp. 6-12. I am slightly embarrassed that the focus in this article is on the Vatican to the exclusion of the people of God. But in the matter of sexual ethics, the official line is what causes the trouble—and the problems here illustrate how little collegial the church is in practice, notwithstanding the rhetoric of the Second Vatican Council to the contrary. In fact, thanks to electronic communications, the church has never been so centralized and clerical as it is today.

9. "Statement of H.E. Archbishop Renato R. Martino, Apostolic Nuncio, Head of the Holy See Delegation to the United Nations Conference on Environment and Development," released 4 June 1992, pp. 3-4.

10. See Paula DiPerna, Jacqueline Pitanguy, Rosalind Petchesky, and Sharon L. Camp, "Population, the Environment, and Women's Rights: Three Perspectives," *Conscience*, vol. 14, no. 3 (Autumn 1993), pp. 2-12. Cf. in same issue, Nahid Touba, "Just Another Member of the World: An African Physician's View of the US Role in Population Affairs," pp. 13-14. Also Betsy Hartman, "Bankers, Babies, and Bangladesh," *The Progressive* (September 1990), pp. 18-21.

11. See "Notes for Speakers on Environment and Development" (New York: U.N. Department of Public Information, July 1991).

12. See Jose Barzelatto, "Abortion and Its Related Problems," in *Infertility: Male and Female* 4, ed. S. S. Ratnam and E. S. Teoh (Park Ridge, N.J.: Parthenon, 1987), pp. 1-8.

13. See Lisa Sowle Cahill, "Abortion, Sex and Gender: The Church's Public Voice," *America*, vol. 168, no. 18 (May 22, 1993), pp. 6-11, esp. p. 8.

5

A Revolution
in Human Ecology

MARY ROSERA JOYCE

How do we live in the world as a home? By learning to be at home within ourselves. But the quality of this inner life depends upon our sense of ourselves as human beings. What do we think it means to be human?

Not a Rational Animal

Aristotle, the philosophical backbone of Western civilization, saw a human being as a sentient organism with the ability to reason. He defined this entity as a rational animal. The Hebrew-Christian scriptures see a human being as a person created in the image and likeness of God. The Bible does not see us as rational animals at all.

Was Aristotle wrong? His frame of reference, his ecological perspective, was basically biological. In that context we are definable as rational animals. But is our context biological? Is our human nature basically that of a sentient organism? Or are we basically persons who are like God?

Speaking logically, in what genus are we? In the biological genus of animals along with foxes, wolves, lions, and tigers? Or in the ontological genus of persons, along with the angels, the Father, Word, and Holy Spirit?

First of all, we are much more intuitive than we are rational. All persons are intuitive; only the human person reasons from one judgment to another. Intuition is a direct form of knowing which is infinitely clear in God, finitely clear in angels, and sometimes clear, but most of the time cloudy, in human beings. Intellectual intu-

47

ition in humans has dawns, noon-days, dusks, and moonlit nights. Nevertheless, we are probably about 90 percent intellectually intuitive and only about 10 percent rational. Because we are so predominantly intuitive, we actually know far more than we know that we know. Reason clarifies what we already know. Reason is an explicator of intuition. But reason is not our original capacity for intellectual knowing.

Rationality in us is like the smallest part of an island above the surface of a body of water. We are basically intuitive, bodily persons, not rational animals. Every power within us is the power of a person. Nothing within us, not even our eating and elimination, is the power of an animal, even though some of our powers are animal-like.

A Bodily Person

What is a person? Basically, a person is a being with intellectual, volitional, and affectional powers. A person is centered not in the intellect, but in the heart, especially in the affectional power of the heart, In a person, the intellect guides and serves the power to love. The affectional depth of the heart needs the light of intuitive intellection as the source of wisdom, and it needs the clarifying power of reason. But in a person, not in a rational animal, love is primary.

A person is one who loves, and one who knows in order to love. We say along with John the Evangelist, "God is Love," not "God is Knowledge." Created in the likeness of God, we are made to love above all else. We are destined for beatific love, as well as for beatific vision, but primarily for beatific love, which beatific vision increases and enhances. Vision serves love here on earth and also in heaven. Wisdom without love is almost nothing at all.

Our nature is not that of persons among animals nearly as much as it is that of one kind of person among other kinds of persons. We are human persons among angelic and divine persons. When we see our nature clearly, in its true ecological perspective, then we can begin to understand how we are to live in this pre-paradisal universe.

Are we here to dominate and control nature? Is dominion the same as domination? Rational animals dominate, manipulate, and control nature. Persons have dominion over nature without domination. Persons understand their freedom differently than do rational animals.

Human Freedom

True human freedom does not mean, primarily, the ability to control ourselves or anything else. Control is secondary within our freedom. Controlling is masculine. Freedom is basically feminine, basically receptive, not controlling. Control is authentic only when receptivity precedes it. Then the feminine and the masculine (receptivity and a healthy kind of control) can balance each other in harmony.

Freedom is basically our ability to receive the gift of being as it is given to us. Receptivity does not exist in animals—nor in rational animals. Receptivity exists deep in a *person's* prerational life and inner space between thoughts and actions. Because of inner receptivity, persons know how to have dominion over nature without domination.

The Human Difference

The human being differs from other persons by being a bodily person. But *bodily* does not mean "embodied," as if the person inhabits a body. Western as well as Eastern philosophy is infected with the disease of embodiment, as if the body is something that is done to a soul. We are bodily to the core, to the depths of our soul and of our personhood.

Humans are like animals in some ways, but humans are not animals in any sense at all. Every biological structure or process in a bodily person, such as eating and sleeping, is personal, that is, of a person, not of an animal. That makes a critical difference between animals and us, no matter how much we are like animals. Reason is not the difference. Human personhood is.

We are more like God than like the animals. How, then, does God live in this world? How did God, as the Son of Man, live in this world? With loving dominion in his carpenter's shop. With loving appreciation for the lilies of the field and the birds of the air. With love for all persons, especially for sinners. With a readiness to forgive, and bless, and heal. With a good catch of fish, and a multiplying of the loaves. The Father so loved the world that he gave his only-begotten Son. Do we so love the world that we give our only-begotten consciousness with an attitude of receiving the gift?

Human Sexuality

If man and woman together are a likeness of the God Who Is Love, what does this mean about their sexuality? A revolution in human ecology also becomes a true sexual revolution. The false sexual revolution of the Freudian irrational animal has produced so many human ecological disasters (including the plagues of sexually transmitted diseases, AIDS, and abortion) that we should be looking, now, for a true revolution.

The sexuality of a rational animal is 100 percent genital. Though the discoverer of the irrational animal, Sigmund Freud, named oral, anal, and phallic, as well as genital stages of sexual development, he still saw genital sexuality as an animal function and as the reason for all other sexual functions.

Our sexuality, created in the image and likeness of God Who Is Love, is not basically genital at all. Genitality is only about 10 percent of the sexuality of a human person. When we see ourselves basically as bodily persons in the likeness of God, we see that Jesus, the Son of God, a perfect celibate, was sexually perfect and sexually fulfilled. We see ourselves in his revelation of our nature. We can be sexually perfect and sexually fulfilled with or without genital intercourse and procreation. We are, indeed, sexually free. Biology courses treat sex as a physical necessity along with the need for food and water. This biologistic view has devastated the self-concept of human beings.

Intimacy and Co-creativity

What is the sexuality of a person that is so different from that of an animal or a rational animal? The sexuality of a person is a differentiation in being of receptive and expressive reciprocity that moves a being toward intimacy and co-creativity. The receptive side is feminine, and the expressive side is masculine. The receptive and the expressive together create intimacy. Together they co-create. Genital intercourse is only one form of sexual intimacy. And procreation is only one form of sexual co-creation.

Are animals capable of sexual intimacy and co-creativity? No. Is God capable of intimacy and co-creativity? Supremely so. We might not want to call this intimacy and co-creativity sexual in God. But whatever it is, man and woman, and their intimate, co-creative relationship is a special likeness of intimacy and co-creativity among the different persons of God. And the Divine Persons are different. For example, only the Second Person became human.

Seeing more clearly our likeness to God, we can begin to understand more fully why our sexuality is so predominantly psychological and spiritual. Intimacy and co-creation between a man and woman happen in everything they share, not just in genital intimacy and procreation. Opening up the immense world of man-woman intimacy and co-creativity in our bodily personhood and sexuality is a revolutionary challenge before us. This challenge is a call to sexual friendship: the friendship of man and woman.

Man-Woman Friendship

The ancient Greeks named three kinds of love: *eros* (desire), *agape* (giving), and *philia* (friendship, mutual sharing). We understand *eros* and *agape* between a man and woman. But we do not understand *philia* very well. After two thousand years of Christianity, we still have barely begun to elucidate the sexuality of persons that makes possible the regaining of the original Adam-Eve friendship, which was lost.

When a man and woman become friends, neither one of them dominates or controls the other. Together they have dominion (not domination) over themselves and the earth. Dominion begins in receiving the gift. And receptivity is not passive. Listening, for example, is receptive and active. True listening is just as active as speaking, though in a different way.

The sexual receptivity of a man is not the same as that of a woman. The sexual expressivity of a woman is not the same as that of a man. The feminine in a man is a masculine kind of feminine, nuanced differently from the feminine in a woman. And the masculine in a woman is nuanced in a feminine way. The two sides of inner sexuality are emphasized or underlined differently in a man and a woman. But receptivity, not control, is basic within the true sexual intimacy and co-creation of man-woman friends.

Sexually Active: Three Ways

Between man and woman, there are really three ways of being sexually active, not just one as in animals. First, there is the inner way of receiving sexual feelings as energy to develop manhood or womanhood without genital activity. This first way develops a deep sexual freedom as the capacity for man-woman friendship.

Second, there is the intimacy and co-creation of mutual sharing of thoughts, feelings, values, hugs, and other friendship activities without genital involvement. Third, there is the intimacy of marriage, genital intercourse, and the co-creation of procreation.

The most important is the first. In the first way of being sexually active the person develops the centrality of the heart and the vision that serves the heart. The second and third kinds of sexual activity depend for their authenticity on the first kind. That is the true sexual revolution. But it depends on a revolution, first of all, in our self-concept as human beings.

More Fully Human

What is more powerful than our philosophical self-concept except divine grace? And yet grace works with nature. Our self-concept as a rational animal, or as an irrational animal, warps our lives and blocks out grace. The concept of a rational animal is as outmoded, archaic, and regressive as the Ptolemaic picture of the universe after Copernicus showed that the earth revolves around the sun, rather than the sun around the earth. We are bodily persons in the image and likeness of God, and this frees us to be friends of each other, friends of God, and friends of the earth. In friendship from the heart, we are meant to live in this world as our pre-paradisal home.

6

The Theotokos Project

BEATRICE BRUTEAU

Ecological spirituality has to begin from a response to these questions: What is the world? What is its meaning? What is its value? How does it work? Then we can ask, How ought we to believe and to behave in it?

I am answering these first questions by saying that the cosmos is a *Theotokos Project*. That is, it is an enterprise in which a finite system is developed by autopoiesis, through evolution, to the point where its complexity and consciousness make it capable of bearing the values of the Infinite Spirit whose expression it is. In a Catholic framework, this suggests the revelation of the incarnation, conceived on a cosmic scale. The cosmos itself is the Godbearer. The cosmos itself is the embodied word of God.

The Exegete

The Psalmist asked the extracosmic Deity, "What are human beings that you should be mindful of us?" (Ps 8:4). And we, in turn, ask ourselves, "What is the cosmic process, that we should be mindful of it?" It is not enough to reply that it is the condition of our existence, for we are already sufficiently transcendent of our (merely human) selves to yearn for—even demand—something more comprehensive and more awesome. We want to answer in terms of the largest and grandest idea we can conceive and in view of the most profound and powerful motivations we can muster.

It is not just that the cosmic process is God's creation, marvelous though that answer is. In the Christian mythic (sacred expression) framework, the central icon is the descending and ascending dynamism of the incarnation (Jn 3:13). God's presence in the world, as world, is the revelation, and remains the great mystery.

> I came from the Father and have come into the world; again, I am leaving the world and going to the Father. (Jn 16:28)

> And behold, I am with you all the days until this age of the world reaches its completion. (Mt 28:20)

> No one has ever seen God. But the *monogenes theos* [the only-begotten or born-from-one-parent God], the one being in the bosom [the "hollow" or "pocket"] of the Father, that one *exegesato*. (Jn 1:18)

An *exegetes* is a guide, a counsellor, or an interpreter of oracles and sacred mysteries. The Incarnate God is the Exegete, the revealer of the sacred mysteries, which would otherwise remain invisible. The Exegetic Act, as we may call it, comes forth from the invisible Source, from its very hollowness, the transcendent Void of the Godhead; it manifests the Deity as a word spoken reveals the speaker, makes the speaker present and accessible. It shows the invisible—the inconceivable, the formless, the infinite—in the only way the Absolute *can* be shown: as the finite, the formed, the relative:

> Philip said to him, "Lord, show us the Father, and we shall be satisfied." Jesus said to him, "Have I been with you so long, and yet you do not know me, Philip? He who has seen me has seen the Father; how can you say, 'Show us the Father'? Do you not believe that I am in the Father and the Father in me?" (Jn 14:8-10)

The figure of Jesus in the Fourth Gospel is identified with the Word of God, through whom "all things were made" (Jn 1:3). The Word "works" in the world; these works display the wisdom, the power, and the love of God in a process which is growing toward a return to the Father. The Epistle to the Ephesians speaks of "a plan for the fullness of time, to unite all things in him, things in heaven and things on earth" (Eph 1:10). This unity will supersede all local and particular orders and governances, and it will in turn be arrayed under, subsumed into, and totally penetrated by the divine, "so that God may be all in all" (1 Cor 15:24-28; cf. Eph 1:21-23).

The Christic Figure thus seems to be capable of being expanded beyond a single individual human being, even beyond being a community of human beings (the church). It must encompass "all things," both when they are made and when they come to their "end" in the ultimate unity. Therefore I propose to treat of this Exegete as the Cosmos itself, that is, regarding the Cosmos as the "body" in which the Word of God has become incarnate.

Expanding the Christic Exegete to the dimensions of the Cosmos puts a new light on some of the familiar sayings. For instance, the text just quoted, "Have I been with you so long, and yet you do not know me?" takes on a further significance when understood as spoken by the universe. Certainly, it has been with us for a very long time, and we should have known it as the presence of God by now.

Ever since the creation of the world his invisible nature, namely, his eternal power and deity, has been clearly perceived in the things that have been made. (Rom 1:20)

God-the-Revealer is the Child of God-the-Invisible, what has issued from the Hidden One's own interior, birthed by God as God's own self-expression. Therefore, it is typical of this Divine Child to say, "Why did you [think that you had to] search for me? Did you not know that I am bound to be present in all the affairs of the Father?" (Lk 2:49). But many of us do not understand this word the universe speaks to us.

The Cosmic Exegete goes on to say, "I do nothing on my own authority; it is the Father dwelling in me" (Jn 14:10); "I have come down from heaven not to do my own will but the will of the One who sent me" (1 Jn 6:38). And yet, "All that the Father has is mine" (Jn 15:16); "I am in the Father and the Father is in me" (Jn 10:38), "the Father and I are one" (Jn 10:30).

My understanding of the Catholic tradition as supportive of ecological values is that Reality is both the Formless, the infinite invisible Father-Source, and the Formed, the finite phenomenal Word-World. The Formless expresses itself as—gives birth to—the Formed, and the Formed constantly testifies to and proclaims the Formless. To see the Formed—with enlightened eyes—is to see the Formless. The Formless, the Father, is not Something Else, as something beyond, not here, not this. It is really present, fully present, right here, as This. And yet it totally transcends any form and all forms put together: "The Father is greater than I," says the Word (Jn 14:28). This is the mystery of the incarnation.

The incarnational exegesis of the Invisible comes to fulfillment when the Cosmic Exegete knows itself as the Offspring of the Source, when it grows up to consciousness and is able to "bear witness to [itself]," saying "I know whence I have come and whither I am going," and "If you knew me, you would know my Father also." This realization has to be clear and strong: "I know him. If I said, I do not know him, I should be a liar . . . but I do know him and I keep his word" (Jn 8:14,15,55). When the Cosmos has come to such consciousness, we may recognize that third presence of the divine that we call the Holy Spirit, the Spirit of Truth that gradually—for it cannot all be borne at once—reveals all the truth (Jn 16:12-13). It is the bond of unity, tying back the expressed Word-World to the generating Father-Source.

This is the beginning of an answer to the questions about the nature and value of the world, the ground for believing in ecological values and practicing ecological virtues. Since all forms are forms of the Formless, children of the Invisible God, this is the foundation for that universal unconditional love which is the identifying mark of the children of God (1 Jn 3:10). Recognizing this deepest of all kinships, we acknowledge all as our relatives and genuinely experience that what happens to any one of them happens to us (Mk 3:35; Mt 25:40).

The Theotokos

Now I want to introduce another icon for the sacred world, one which carries the same message but reverses the metaphoric roles. This is the icon of the Theotokos. The Father/Son icon emphasized the generative character of the Invisible, for which reason it was named Parent. The icon of the Theotokos takes up the birth-giving character of the world and regards the offspring as divine. The Catholic mythic tradition is rich with sacred stories, images, and emblems that can be read on many levels. Now that we are once again unafraid to admit that much of the spiritual power of a religious tradition resides in its multi-layered mythic dimension, we can comfortably talk about the extended and allegorical significances of even the primary figures and doctrines. This is what will be done here with the figure of the Theotokos, the "Godbearer," personified and historically incarnated by the Blessed Virgin Mary. The term came into use in the church's refutation of the Nestorian assertion that there must be two persons in Christ if he has two natures, and that Mary could not be the mother of the divinity. The Council of Ephesus (431 C.E.) reaffirmed the unity of the person of Christ and buttressed it by declaring that his mother, Mary, was truly the Mother of God, the Godbearer. The aim of the argument was to protect the divinity of Christ, in spite of his having come forth from humanity. This is the point of what will be said here in a cosmic context to protect the value of the ecological expression of God's Word, despite its being developed in a space-time-matter-energy universe. In this expansion of the icon, the Cosmos becomes the Theotokos.

The material world as Theotokos is, perhaps, remotely foreshadowed by the Hebrew *Shekhinah* (first used in the Aramaic interpretive translations of the Old Testament called Targums: *shekinta*), meaning the dwelling or presence of God on earth. It was used to avoid a misleading anthropomorphism and to signify some form of divine immanence. In Kabbalistic mysticism, the *Shekhinah* is regarded as the female aspect of God.

The metaphor of the female Godbearer gives us another theme that is important to our ecological considerations, and that is the sense of *gestation*, of gradual formation and emergence. The divine is hidden in the body of the material world, where it grows secretly, in terms of its own internal programs of formation, until it can be birthed into full visibility. A feeling for this value in the world shows in medieval alchemical work, where the metaphor for the development of spiritual consciousness out of human animality is the transmutation of base materials into the gold of the spirit: the liberation of God from matter. Carl Jung felt that "for the alchemist the one primarily in need of redemption is . . . the deity who is . . . sleeping in matter." The alchemist is working to see "whether he can free the divine soul."[1] The alchemist thus becomes a kind of midwife, understanding and assisting the birth, while the actual mother is the cosmos itself, and the child to be born represents divinity incarnate in the world. One thinks of the Epistle to the Romans:

> The whole creation has been groaning in travail together until now. . . . For the creation waits with eager longing for the emergence of the children of God . . . because the creation itself will be set free from its bondage to decay and obtain the glorious liberty of the children of God. (Rom 8:22,19,21)

This aspiration has been taken up by poets:

> The cherub with his flaming sword is hereby commanded to leave his guard at the tree of life, and when he does, the whole creation will be consumed, and appear infinite, and holy, whereas it now appears finite and corrupt. . . .
>
> But first the notion that man has a body distinct from his soul is to be expunged.
>
> If the doors of perception were cleansed every thing would appear to man as it is, infinite.
>
> (William Blake, *The Marriage of Heaven and Hell*)

> . . . At the very worst
> It must have had the purpose from the first
> To produce purpose as the fitter bred:
> We were just purpose coming to a head.
> (Robert Frost, "Accidentally on Purpose")

The gradual appearance of purpose, consciousness, freedom, holiness, divinity, the infinite, is the gestation taking place in the cosmic Theotokos. It is what I will call the Theotokos Project. It is a cosmic effort, work, organic process, growth, whose intention is to bring forth the values of the infinite and the divine in the forms of finitude and matter.

> And Jesus progressed in wisdom and age and giftedness before God and humanity. (Lk 2:52)

The progressive growth of the Divine Child is the story of the universe, the unifying factor that makes it a single intercommunicating system. In this systemic growth, everything participates, everything contributes. Nothing is insignificant, nothing is lost, nothing is wasted, nothing can be hidden or held private. Everything is knitted in and has its consequences. It is a community project. The offspring of the Theotokos is "the image of the invisible God," in which "all things hold together" and in which "all the fullness of God" can "dwell," because "all things, whether on earth or in the heavens" are thereby harmonized and brought into unity (Col 1:15,17,19,20).

The fructifying stage of the Theotokos Project comes when the Cosmos-Community becomes conscious enough to realize that it is the self-expression of the Invisible God. Thus the world does not exist to be escaped, or to be subjected to human manipulation, or to be neglected, but to be nurtured to "maturity, to the measure of the fullness of the stature of Christ," as a single and "whole body, framed accurately together and lifted up together, through every connection and channel of communication and supply, according to the operation in measure of each one's part"; in this way "the growth of the body" advances as something that is "building itself in love" (Eph 4:13,16).

The whole cosmic event has been ecological from its inception, and it must be ecological in its progress and in its culmination. This means that the processes by which the parts of the cosmos are in communication with one another, particularly the constructive and destructive aspects, the coming into being and the passing out of being, are also the divine gestation. They do not exist merely as means to an end. They have a value to be revered even now in the midst of their incomplete process. Indeed, in spite of our speaking in terms of fruition and culmination, we can also say that the value resides in the process as such, or that the process is not to be judged as incomplete, or that the process may never be finished. The cosmic ecological process is, just as it is at any stage of its development, the dwelling of God and deserving of veneration.

The Theotokos Project is a kind of squaring of the circle, in which the square represents the cosmos (in primitive symbology, the earth—four directions; now we might say four dimensions) and the circle represents the divine, the transcendent (primitively, the heavens, equal in all directions; later, that whose center is everywhere and circumference nowhere). The question is whether a square can be constructed with the area of the circle, whether a cosmos can be grown with the value of God's self-expression, while at the same time acknowledging that the infinite and the finite are strictly incommensurable.

> The Father is greater than I. . . . I do not speak on my own authority; but the Father who dwells in me does his works. (Jn 14:28,10)

And yet,

> I and the Father are one. (Jn 10:30)

The Image of the Creator

A related question is whether it would be possible to create a finite universe and conceal the fact of the creation from the creatures—including the Creator's deity, the Deity's creativity, and the fact that what is created must resemble the Creator. If we are going to think of the world under the metaphor of something that is made, then we are bound to include in that metaphor our experience of the act of making, to which we always commit something of ourselves—our ideas in the form of the thing made, perhaps our feelings in the way that the thing is made, and certainly our will in the fact that the thing is given existence at all. This minimum self-revelation cannot be avoided, and usually that which is made carries much more self-expression, especially as the making moves from the merely utilitarian to the purely artistic. In very basic ways, therefore, the created resembles the creator.

In the case of the universe—by which we name all that is in the finite domain—if it is made, it must be made by that which is not in the finite domain. Its creator must be Deity. But how can the finite resemble the infinite? One answer is, by imaging the very character by which the finite is bound to the infinite under this metaphor, namely the creativity of the infinite. That which is made is made in the image of the Creator by being creative in its turn.[2]

When we look for creativity in the world we observe, we see an evolving universe, one that is gradually building itself up, even building itself up in love, by forming compounded unions of diverse items. The universe appears to us at the present time in terms of elementary particles united to form atoms, atoms united to form molecules, molecules united to form cells, cells united in organisms, and organisms united in societies. It presents itself to us as a growing being, advancing in "wisdom"—complexity of form; "age"— through time, each stage building on its predecessors; and "giftedness"—range of behaviors; "before God and humanity"—in relations of interior centeredness and in relations to relatively exterior beings.

The diversity of forms, even on our one small planet, displays creativity on a grand scale: two million species of flora and fauna, with four hundred having lived and died in the past for every one alive today. But the range of behaviors of these species expresses far more creativity in their interactions with and modifications of their environments and in their social relations. And in particular, these behaviors indicate a steadily increasing degree of that interior centeredness that we call consciousness. The more consciousness increases, the more imaginative and inventive become the ways in which creatures adapt to their circumstances and adapt circumstances to themselves. An outstanding advantage is gained by the creature's being able to represent to itself interiorly a proposed course of action and test out its probable consequences by memory of similar occasions. This kind of abstraction and internal processing of practical matters prepares the way for similar fantasy projections and abstract processing that does not relate to practical matters but is of a speculative and artistic character. It is hard to see what limit can be put on the number and kind of creations produced by the conscious creatures in this way. Just counting the possible states of the human brain gives a super-astronomical number.

An idea-processing creature of this degree of complexity is bound to discover the connection between itself and the Creator, between the creativity that produced it and the creativity by which it produces its own creations. It will come to know itself as a continuing image of the original Creativity itself. In a moment of mythic intuition, it will call itself a child of God.

The advanced consciousness-complexity of the child of God will enable it to view the universe as a whole and to view it not just in relation to the viewer but as if seeing it in its own terms. The universe will reveal itself under difference images (for the discerning consciousness will realize that it can only make models for representing the universe's behaviors to itself), one of which is now the system. A system is "a dynamic order of parts and processes . . . in mutual interaction."[3] This is fundamental to our sense of ecology, the living together in one household of many relatives, all of whom make inputs to the common system and all of whom are affected by the others' inputs.

The universe-system will then be understood as something moving, something changing, and the changes will be seen to evolve in the direction of increasing distance from equilibrium of the system, carrying rapidly increasing quantities of information, resulting in increasing flexibility of the subsystems (creatures in the universe)—which means ever greater creativity of the subsystems and of the universal system. The universe will appear to be creating greater scope for its creativity.

Like a living organism, the universe is an intrinsically active system, characterized by growth. An embryo develops according to a pattern within it, modified by—perhaps interrupted by—the conditions of its environment. The community of species undergoes evolution, which in many ways is like development, diversifying and surviving. Is there a more general pattern that underlies the random experiments made by genetic variation, some basic thrust to expansion, accumulation of advantageous characters, increased flexibility, and consciousness? Is the drive toward ever-greater creativity the one thing that is built into the universe—unavoidably built in, because the universe is the work of Creativity, which cannot keep its image from being impressed upon its product?

And when Creativity has created such growth toward increasing creativity, and the creatures have grown to the ability to contribute to the creativity and to know themselves as creative, then has the gestation of the Divine Child in the Theotokos come to a significant stage—shall we say to quickening? to a certain reflexivity and self-possession?

This brings us to the strangest and most wonderful aspect of the universe as made by God in the image of God: it is autopoietic, self-making. Autopoietic systems are "self-renewing, self-repairing, and unity-maintaining autonomous organizations of components capable of interactive linkages."[4] The components affect one another and respond to one another in an ongoing conversation. The result of this conversation is that the component processes renew themselves, rebuild themselves, maintain their network, their pathways of interaction, the intercommunication that binds them together into a unity. And they extend themselves, they change themselves, they improve themselves, they transcend themselves.

Is this an image of God? Scripture says, "For as the Father has life in himself, so he has granted the Son also to have life in himself" (Jn 5:26). Philosophers have often referred to God as *causa sui*, and theologians have worked out models of the Trinity in which processions, generations, and spirations figure as processes by which the Persons are interactive and might be said to maintain their unity by their intercommunion. Does Deity also transcend itself? How can the Absolute, the Infinite, transcend itself? St. Augustine felt it could not, and that the only way a temporal world can image an infinite and eternal God is by constantly changing. But, perhaps, self-transcendence in God shows precisely as the act of creation of the finite? I like to interpret Philippians 2:6-7 this way: God does not cling to being only God-the-Formless, but by self-emptying (self-transcendence) is able to take on *form as such* and be born in the likeness of a universe.

Is the making of the cosmos, then, an autopoietic or an *allopoietic* deed? An allopoietic operation produces something different from itself. It seems to me that we shall have to answer, both. Clearly the mystery of the incarnation and the whole ambition of the realization of being children of God indicate that some marvelous joining of these two ways is being pointed to. Image-making itself is ambiguous; when you make an image of yourself, are you making something strictly other than yourself, or a kind of extension of, or incarnation of yourself? What about works of art? What about reproducing ourselves in children? We say the artist lives and is present in the work of art, and also that the art work takes on a life of its own. We are concerned that children should be recognized as persons in their own right, not extensions of their parents, and yet life is a continuous stream, in some ways a single phenomenon including all species.

It is in such questions that the value of the cosmos is lodged. This is what the original quarrel over the Theotokos was about. Is the Incarnate Word such a union of the divine and the human that the human truly can be said to be the Mother of God? In our question, Can God make something other than God that nevertheless can carry, bear, reproduce the characteristics and values of God— as for instance, by being creative, even self-creative and self-transcending? by being intercommunicating and thereby one? And if this should be so, what respect, veneration, and care do we owe to the Universe? To spell this out just a little further, before drawing conclusions about how we might believe and behave, I want to explore the Theotokos Project under the aspects of experiment, art work, and love-gift.

The Theotokos Project as Experiment

There is a sense in which the Theotokos Project is an experiment. It takes place in real time, that is, with an unknown, not yet existent, future. It takes place in real finitude, composed of the relations of the component features of the cosmos to one another. It is not determined in all its details, or in the patterns formed by its processive components, although it has lines of determinism running through it. The more complex it becomes as the local systems advance from physical to chemical to biological to social to psychological to moral to spiritual, the more randomness, undeterminedness, unpredictability, and freedom must be counted as significant characteristics of the systems. This means that the future—"how it will all turn out"—is genuinely *unknowable*, by God or anyone else. Thus the creative evolution of the world may be said to be a real experiment on God's part. A Jewish legend has God saying to the angels as the universe is started on its way, "Let's hope it works."

This idea of regarding the universe as a Theotokos Project was provoked by the claim of Edward Fredkin that the universe is a computer—in particular, a cellular automaton.[5] (A cellular automaton is composed of any kind of units that have the property of being in one of two states—on or off—at every instant of time; whether a given unit is on or off is determined by a rule connecting it with the states of its neighbors in the preceding instant; the rule is called an *algorithm*.) One of the points of interest was that the patterns produced by this process are necessarily unpredictable. The system is so complex (although basically so simple) that it cannot be modeled by any system less complex than itself. This means that it actually has to run in real time to see what the pattern will be at any point in the future; you can't represent the process by an equation and simply solve it for that moment in time.

Two further points of interest appeared: One, if the universe is a cellular automaton governed by an algorithm, what is that algorithm? And two, if the universe is a running computer, what is the question whose answer is developing in it? Here my reaction skipped to a metaphorical plane, relative to Fredkin's physical theory, and I answer that the algorithm is given in the first chapter of Genesis, "Be fruitful and multiply," or, more generally, "Be more!" Be more numerous, more diverse, more related, with greater ranges and diversities of behaviors, with more consciousness, more freedom, and more beauty. This is, of course, not a deterministic rule, but it is somewhat recursive in the sense that the "more" has reference to the preceding state of the universe.

As to the question whose answer is developing, I propose: Is it possible to create a finite world which will become so complex and so conscious that it can bear the values of the infinite? To quote Genesis again, Can God indeed "make man in our image, after our likeness" (Gn 1:26)? Can God make a divine incarnation, a Theotokos? This means that the algorithm must be not only "Be more," but "Be more like God." Such a suggestion inevitably brings to mind the prayer of Christ in the Fourth Gospel, "that they may be one even as we are one" (Jn 17:24), an aspect of the likeness that will come up again.

If this is a real question, and if it is being "answered" through the "running" of the universe-computer, then the universe and the running become of inestimable value and our behavior with respect to the running exceedingly important. Also, if it is true that the answer cannot be projected but can be obtained only by letting the experiment go forward, then a number of remarks can follow. For one thing, if the outcome is unknown, then it is not only unknown for what we presently regard as the future, but it was unknown also at every stage of development in the past. Does this mean, for instance, that the possibility of feedback was unknown before it happened? that sensitivity was unknown, and therefore that pain and suffering were unknown? Is it true that these things cannot be known by analysis of finitude as such, but must be learned through experience?

> In the days of his flesh, Jesus . . . although he was a Son . . . learned obedience through what he suffered. (Heb 5:7-8)

The author of this text goes on to say, "About this we have much to say which is hard to explain," and that certainly bears repeating in the context into which we have translated this saying.

Learning may be absolutely of the essence of the whole Theotokos Project. When we call the universe a computer, we may be thinking of a computer that is completely programmed, line by line, and absolutely determined in the process it can perform. But we now have connection machines, which have to learn how to perform their processes.[6] In these machines the path of the developing answer is not programmed but must be devised by the instrument itself, guided by the feedback that it gets from the interaction of any proffered answer with the appropriate environment. Our universe may be much more like this; indeed, it may be like a connection machine which keeps creating new possibilities for connections. If the universe-computer is designing itself as it goes along, then things are much more complex and unknowable than even Fredkin had supposed. Beyond the very general "Be like Us," there is no information as to the will of God with respect to what we should do. Creativity runs through the whole enterprise, from stage to stage, from moment to moment.

Information is not so much what we start with as what the universe generates as it develops. The configuration of local areas and their relations to one another all constitute information, more and more unlikely—unexpected—arrangements of perceptibles. All their processes and their interactions are information communications. If we attempt to analyze beings reductively, we again find that they must be described in terms of information relations—all the way down. As Fredkin says, everything in the universe seems to be made of information, and information itself is not made of something else, but just information again.[7] So perhaps the universe is not so much like the computer, in the sense of the hardware, as it is like the software. Already in 1948 James Jeans was saying:

> The concepts which now prove to be fundamental to our understanding of nature . . . seem to my mind to be structures of pure thought. . . . The universe begins to look more like a great thought than like a great machine.[8]

Another way of saying this is that the universe consists of acts of communication. This is a very Catholic idea. Creation itself and everything that takes place in the created world is a matter of the communication of being: of existence, of form, of consciousness, of value. The ground of being is a Trinity of Divine Persons, a holy community, in whose image the conscious creatures at least are to unite themselves in a corresponding holy community, the church, whose central sacrament—sign of its interior life—is called holy communion. We may extend this insight by regarding all the creatures of the universe as participating in a sacred communion, sharing their beings with one another.

This sacred communion, this being-sharing of all the systems and systems of systems of which the cosmos is composed, is what we mean by ecology. It is the ecology that undergoes evolution and complexification and increasing consciousness. It is the ecology that practices creativity with respect to its own processes. Ecology is, in fact, the name of the whole game. It is what is going on. Everything that happens is a subprocess in the universal ecology. The Ecology is the Theotokos.

The Ecology-Theotokos is experimental. It runs in real time, creatively, unpredictably. As a Christ-growing, it is historical. Catholicism has always insisted on the historicity of its sacred events. We may, perhaps, say that the historicity of the Christ-Event is an icon of its own larger meaning: Christ comes in history in order to say that history is how Christ, Child of the Theotokos, comes. Thus it cannot help saying that "it has not yet appeared what we shall be" (1 Jn 3:2). The experiment is still running.

The Theotokos Project as Art Work

The suggestion that the universe is made of communication reminds us of the traditional attention given to God's Word, the self-expression of Deity that is responsible for the existence and form and coherence of the world. This means that an apt metaphor for the creation is that it is an *art work* on the part of the Creator. Ephesians 2:10 says that we are God's *poiema*, God's poem. Poems are mediators between the visible and the invisible. Rainer Maria Rilke described the task of the human poet:

> We wildly collect the honey of the visible, to store it in the great golden hive of the invisible . . . the work of the continual conversion of the beloved visible and tangible world into the invisible vibrations . . . of our own nature. . . . It is our task to imprint this temporary, perishable earth into ourselves so deeply, so painfully and passionately, that its essence can rise again, "invisibly," inside us.[9]

I suggest that God's act of creating a poem is the reciprocal of this. God is converting the invisible vibrations of the divine into the visible and tangible forms of this world, with a passionate commitment so that through the painful process honey can be made and the essence of divinity rise again, visibly, outside the Deity.

It is again a conjunction of the infinite and the finite, again a making of a Theotokos. To call it art work is to point to the self-expressive quality of it, the personal presence in it, the intention, desire, ecstacy that are given body by this act. The category of art enables us, Joseph Campbell says, to recognize "all things . . . as epiphanies of the rapture of being."[10] The well-known fact that mystics usually resort to poetry in an effort to express their experiences suggests that we may think of God's poem-making as a similar effort to express the inexpressible. It further suggests that perhaps we should not read the poem too literally, that we should expect many levels of meaning in it conveyed by quite distinct methods (as in a word-poem the artistic effect is achieved also by rhythm, melodic sound, rhyming, image association, formal structure, and so on). Franklin Merrell-Wolff remarks that "when the 'Voice of the Silence' speaks into the relative world, the Meaning lies between the words, as it were, rather than in the direct content of the words themselves. The result is that the external meaning of the words, in a greater or less degree, seems like 'foolishness,' as St. Paul said."[11]

This approach to the world is an antidote to the contemporary attitude to the environment as a utility. This view says that *usefulness* is not the point, is not the value concerned. An art work is not a value addressed to some other end, is not in service to something beyond itself, is not a means or an instrument. An art work exists for itself, is its own end and value, in the space between the infinite and the finite, and in the communication-space between the creator and the appreciator. The name of this value is *the aesthetic*, an irreducible category of significance. More simply called *beauty*, it is recognized in Catholic philosophy as one of the transcendentals, a value attaching to everything that exists.

The metaphor of the poem brings out another point for our appreciation of the world as God's artistic epiphany. The poem is not beautiful only when it has come to an end; it is beautiful in its transience, in its motion, in its development. One does not read a poem, sing a song, perform a dance, in order to get it done, to finish it. It exists in its beauty only while it is being done, precisely in the doing itself. The implication of this view is that the world, although it is historical and develops in time, is not trying to come to the end of time in order to attain its full value. It is carrying the divine presence all the while. An appropriate gospel story seems to be the conversation between Jesus and Martha just prior to the raising of Lazarus. Martha acknowledges that her brother will rise again "at the end of the world." Jesus rejects and corrects this view, declaring "I AM the resurrection and the life" (Jn 11:25). This strong assertion of the present tense, with its clear allusion to the name of God, the Eternal, reframes all existence as the presence, the expression of the Eternal in and as time: time-as-present bearing the presence of the Eternal.

If we see the Theotokos-Ecology in this light, as God's art work, not utilitarian, not goal-seeking, existing and being beautiful in its own right and for its own sake, we have to revise many of our attitudes and actions. But the satisfaction and sense of meaningfulness in our existence may quite transcend what we could experience in a utilitarian context. Instead of regarding the material world as something we may first despise and then exploit, we are called to reverence it and delight in it. Recognizing in it the extended version of the human Theotokos, shall we not sing to it also, "*Tota pulchra es*"—You are all beautiful and there is no spot in you. Matter is not the source of evil but the art medium of God's self-expression, the matrix in which the divine aesthetic is being manifested.

The Theotokos Project as Love-Gift

To find a vessel strong enough to "bear the beams of love," as William Blake puts it, is the mystery of the incarnation. The Theotokos Project is the making of that vessel. It is a vessel for love, made by love. Catholicism bases itself on the revealed announcement that "God is love" (1 Jn 4:81). Everything done by God is done out of love, is an expression of the love that unites the Persons of God in the Holy Trinity. The cosmos, therefore, is a love-expression, a love-gift.

The love that God is said to be is *agape*, the love that is distinguished from our ordinary self-seeking loves by being devoted entirely to the perfection of the beloved. Thus it is totally gift, gift of self, gift of being. The act of creating the world, in the Catholic view, is an act of pure love. It gives the universe its existence and continues to nurture it to develop it as the image of the divine love, as the vessel that will bear the beams of love, as the Theotokos.

The first thing one notices about this originating, being-giving love, is that it is *unconditional*. It is not a response to a value perceived in the beloved. It is not a reaction but a first action. It is not provoked by or dependent on anything other than itself. There are no conditions that have to be satisfied before it acts. The question of deserving it cannot even arise. It is absolutely impartial. It does not withhold anything but gives itself without stint.

Can a vessel be found to bear such excessive beams? Who can endure to be so loved? Who can accept the call to love in such wise in turn? The hoped for answer is—the universe. In our extended interpretation of the Theotokos, it is the universe that is hailed as highly favored, full of grace, a blessed matrix, in which will develop the image—the offspring—of the love that created it.

The second thing one notices about agape is that by this one principle both differentiation and union are established. Usually we require one principle by which to differentiate and another by which to unite, but this love does both. Inasmuch as it is *gift*, it relates to the *other* as such, the not-self, even granting being to the other if necessary. Inasmuch as it is *self*-gift, it unites with the other.

Joseph Campbell, following James Joyce, makes much of a mysterious verse in the Epistle to the Romans: "For God has consigned all men to disobedience that he may have mercy upon all" (Rom 11:32).[12] It is related to equally strange words in the *Exsultet*, the song to the Paschal candle in the Roman liturgy of the Easter Vigil. The text contains acclamations of the "truly necessary sin of Adam . . . blotted out by the death of Christ," and the "happy fault, which deserved to possess such and so great a Redeemer."

O truly blessed night, which alone deserved to know the time and hour in which Christ rose again from the grave! This is the night of which it is written: "And the night shall be as light as the day; and the night is my light in my enjoyments."

Campbell proposes interpreting these lines in a cosmic context that is generic for many religious traditions, the basic idea of which is that the sin and disobedience themes are mythic pointers to the conditions of finitude and creation. This separation from God is necessary in order to produce the creation at all and is the prerequisite to the reunion in which the conscious creature realizes itself as the child of God. This is a kind of love-dance in which first the differentiation takes place and then the union.

This may be an instance of the very interesting case in which a literature and a tradition offer a *moral* theme and storyline as a mythic representation of a still deeper level of reality, in this case, the love-gift metaphysics of the Theotokos. The making of the Theotokos is figured as, first, the "happy fault," the splitting or separation by creation of the world-matrix from the Creator. This separation is worth doing, the myth says, because the reunion to come will be so great. It is a necessary sin, because separation of some sort must be reconciled in the final love union. The unconditionality of the love-gift is indicated in the reference to mercy.

The rising again acquires another overtone now that we have read Rilke's description of poetry and converted it to our own use in the notion that the invisible values of the divine are hidden in the world as yeast in dough; they rise again in the visibility of the Incarnate Word, born of the Theotokos. When this consciousness appears in the universe is a mystery, just when it happens that the world begins to realize who and what it really is. But when it happens, then the night of separated-from-God unconsciousness of matter is in its own terms "light as the day," the divine life values have been made flesh in the finitude of this universe. This is the cosmic love-union.

Since love is a unification of life, it presupposes division, a development of life, a developed many-sidedness of life. The more variegated the manifestation in which life is alive, the more the places in which it can be unified; the more the places in which it can sense itself, the deeper does love become.[13]

A third thing that one notices about agape is that this love-gift, love-dance, produces joy. This joy of life, the life worth living for its own sake, comes from conscious participation in this love-play, rejoicing in diversity and variation, rejoicing in life-sharing and union. "And the night is my light in my enjoyments." The Theotokos is the site of these diversities and these unions. The Theotokos is the "vessel strong enough" to bear the divine love, the "Vessel of honor, Spiritual vessel, Singular vessel of devotion," and the "Cause of our joy" (Litany of the Blessed Virgin Mary).

When the Theotokos is understood as the Cosmic Ecology, we know that we are all members of it, all processes in the universal love, diversifying and reuniting. To realize the joy of it, we have to practice the agape, the unconditional self-giving love. George Bernard Shaw said: "This is the true joy in life: the being used for a purpose recognized by yourself as a mighty one, the being thoroughly worn out before you're thrown on the scrap heap; the being a force of nature instead of a feverish little clod of ailments and grievances complaining that the world will not devote itself to making you happy."[14]

The hope of the cosmos to become a true Godbearer makes all our lives meaningful and worthwhile as participants in and contributors to this great enterprise. The indeterminate nature of the work, the unpredictability, the constant coming of the unexpected, is something we have to learn to accept and to treasure. It is essential to the success of the project. We cannot be protected against all accident or failure, against all sin and suffering. As members of the embodied God, the suffering God, we commit ourselves to the project, risking unknown calamities, for the sake of "the joy set before" us (Heb 12:2).

Believing and Behaving

The question of how we ought to play our roles in this divine experiment, work of art, act of love, can find general answers in this context. Removing the question from the utility context makes an enormous change right at the beginning. Seeing the ecology not as one consideration—and perhaps a marginal one at that—but as the whole of our situation in life, reorients our perspective radically. Seeing the ecology as something sacred focuses our vision more specifically yet. And regarding the sacred ecology as a Theotokos gives us a definite model in terms of which we can reschedule our views and our actions. It further gives us a tie to the tradition which vitalizes the model and expands our understanding of the tradition.

There are several principles, or points for reflection, that can be derived from this view: the sense of the Whole, the extension of the acknowledgment of rights to many others, and the evolution of our sense of evil when these rights are transgressed; the importance of staying open for new developments and variety and being able to live with unpredictability recognizing that the ecology functions as a system rather than like an empire; and being patient and faithful as we struggle toward the birth we expect—actually a kind of ongoing birth, or ever greater birth.

The basic principle in any sense of ecology is the perception of the Whole as the foundational reality. Differentiation takes place within the whole, and the more creative the differentiation, the more profound the unity and the wider the scope of the activity. The distinct contributions of the various members need to be protected by the other members if the Whole is to flourish and each member thereby enhanced. Thus simplistic self-interest gives way to complex community interest, self-centeredness to all-together centeredness, in which the sharing activities of the many participants are what make the Whole work, make it—them all—live and develop and enjoy. Motivations make a subtle shift from desiring that life be good for me to desiring that life be good for us; on the active side the desire that I should achieve the good expands into the will that the good itself be attained. There is a growing realization that the good cannot be attained by parasitic activities, in which one member is served by others or others are seen as existing for the convenience of the privileged one. Such privilege begins to be seen as unhealthy. Relations need to be symbiotic, enhancing the general life, the shared life.

This sense of the Whole is the way the ecology as Theotokos begins to bring forth the value we now discern as divine: in a Catholic context, the value of the Trinitarian Life. Unity, with internal distinctions or differentiation that bind the unity as well as derive from it, is the way we think of the Godhead. If what the Godhead is undertaking in creating the Theotokos is to make an image of itself, then this kind of wholeness is a fundamental part of that effort.

In order to make a finite universe, it is necessary to make different kinds of things and to enable them to maintain their distinctions. Only in this way can they relate to one another to make even greater diversity of compounds. What we, later, in the moral era of the universe, call self-seeking—protecting and promoting oneself at the expense of one's environment—has been the natural law of a universe that is working at development. Dissipative processes build up high degrees of local order and complexity by sucking in appropriate raw materials, which they then organize into themselves, and by discharging what from their point of view is waste. To a certain extent ecologies operate on market dynamics, and to a certain extent symbioses form in terms of enlightened self-interest. There also seems to be, among the higher forms of life, some appearance of what we recognize as altruism, concern for another without any obvious advantage to oneself.

As consciousness begins to emerge in human societies, a new sense appears: an acknowledgment that (at least some) other members of the community have certain rights—to life, freedom, property, and so on. Respect for these is felt to be an obligation. Disregard of the agreed upon rights is moral evil. At this point the universal evolution adds a new layer of development. There begins an evolution of the sense of evil. What constitutes rights and who has what rights becomes an area subject to expansion (and occasionally contraction). Gradually relations and acts that had formerly been merely natural and perhaps even admirable—for instance, raiding one's neighbors for useful goods and probably killing some of them—begin to be perceived as wrong.

It is a further development and extension of this growing sense that is active now in our concerns about the ecology. Rights have spread from elders or warriors to opposite numbers with whom there are alliances; very reluctantly to other nations or races that had been fair game for enslavement or incorporation into one's empire, even more reluctantly to women, children, and animals, all of whose cases are still in process at the present time. That the ecology as such, or any particular members of it other than ourselves, should actually have rights is the new idea on the horizon, and it is in fierce conflict with the older drives of self-interest that are still functioning in the finite universe.

The vision of a sacred ecology, one that is a potential Theotokos, can make a difference in this struggle among points of view. Instead of there being an environment for us, we perceive all of us together as the subject of interest; this whole being is what has a right to live, develop, enjoy existence. Nothing in this community exists merely as a natural resource for others. Everything begins to be perceived sympathetically, that is, from its own point of view. The animals have a right to life in and for themselves; they are not simply the unprocessed state of the food on our tables. The forests are not only the source of oxygen and wild plants, from which we may extract beneficial chemicals, but have a life of their own to which they have a right. We don't hold off killing endangered species so that our grandchildren will have the opportunity to see rare and interesting creatures but for their own sakes.

This is the sort of consciousness that is arising. It inevitably runs into contradictions, paradoxes, and what still seem to us to be absurdities. The creative resolution of these puzzles has not yet appeared. But the image of the Theotokos gives a perspective that can help by strengthening the confidence that somehow the Whole is of great value, that it is moving toward ever greater value, and that we are called to exercise creativity within it for its whole sake—which includes us but is not limited to us.

We are in the midst of the creative process. The image of the Theotokos as art work especially speaks to us here. The evolving Theotokos seems to be something like an improvisation. It is not following preordained outlines. It is striving to be like God, but then that God is called living and is perceived as Creator. So the Theotokos is groping its way toward being ever more living and ever more creative. This means for us—for our believing and our behaving—that we must be open to novelty, to new views, to paradigm shifts, to broader perspectives, to revisions of our beliefs and attitudes. We cannot settle down with the ways of the ancestors and devote ourselves to preserving our traditions. The only tradition we may dare to keep is the one that says we must be ready to change and grow. We are growing up in the image of a God who says, "Behold, I make all things new!" (Rv 21:5).

> Increasingly we must learn to *live with disturbance, live with variability, and live with uncertainties.* Those are the ingredients for persistence.[15]

We are coming to understand that ecologies are not simply "balanced."[16] Ecologies, like everything living, are always at least a little off balance, as the various components pursue their respective agendas. There is no preestablished harmony laid on by nature—or even by the Creator—but the patterns of interactions arise from the grass roots. This is the secret of the creativity they manifest. Not having top-down control gives the system the flexibility that enables it to adjust to changing conditions and to invent new interactions.

When we participate consciously and deliberately in these systems, then, perhaps the most important thing we can do is *avoid imposing a top-down control.* This, unfortunately, is a principle that is not strikingly in evidence in the Catholic tradition, although it can be strongly argued that it was pointed out by Christ as the way to grow toward the type of unity lived by the Trinity. In his dramatic gestures of Holy Thursday, Jesus rejects the categories of servant and lord and insists on friend—implying equality and life-sharing—as the mode of participating in his Way, Truth, and Life.[17]

But, since almost all of our contributions to the ecology in which we live will be deliberate rather than spontaneous and unconscious, we will inevitably be engaged in some degree of management. Therefore, after first undertaking to do no harm, we must devote ourselves to studying how these systems work. As they are chaotic phenomena, they will not yield simple answers, but a great deal can nevertheless be learned. There is no reason why our own creative enterprises should not work to the benefit of the Whole. Just as the image of the ecology as Theotokos shows us that matter is not the source of evil, so it can help us to see that human technology need not be regarded as an enemy. To relieve the problems engendered by current technology, we need more and better technology. We need to find ways in which all of us—the whole ecology—can live together in reasonable health and freedom. We have to feel our way toward this ideal, for there are no answers coming down from the top; we must improvise and create and test and modify in a truly living, experimental, artistic way. The Theotokos says to us that this is all part of the project itself and can be done in terms of love, of outgoing concern for all the creatures whose lives are shared together.

We must not fear or turn against our own creativity, derived from the divine Spirit and born of the cosmic Theotokos, but nurture and train it to its proper role in the continuing project. Matthew Fox, quoting Meister Eckhart as saying, "The seed of God is in us. . . . The seed of God grows into God," comments:

> The fact that we "grow into God," that we are ourselves part of the cosmogenesis and its patient and evolutionary ways, is attested to by Paul as well. "And we, with our unveiled faces reflecting like mirrors the glory of the Lord, all grow brighter and brighter as we are turned into the image that we reflect" . . . [which] consists in our growing more and more brightly into birthers and creators like God.[18]

Creativity makes mistakes; creativity sometimes fails; creativity experiences suffering in its process. Creativity also does the unexpected and makes extraordinary beauty appear out of the most despised materials. The Catholic acceptance of the cross as part of the Way is a reminder of the paradoxes in the gestation of this immense child of God.

Therefore, we must not grow weary or faint-hearted but persevere, keeping the hope of glory alive and strong, even as we open ourselves to the unknown and acknowledge that salvation—the emergence of great beauty and goodness—may come about in strange ways and by meandering paths. The image of the Theotokos, garbed in light and crowned with stars, can be like a beacon before us, the sign of the marvelous universe in its divine pregnancy. If we commit ourselves to the experiment, the creation, the love which she embodies, we will be able to go forward gladly, full of patience and faith.

Notes

1. Carl Jung, *Psychology and Alchemy* (New York: Pantheon, 1953), pp. 299-300.

2. Cf. Franklin Merrell-Wolff, *Pathways through to Space* (New York: Julian, 1973), p. 109.

3. Ludwig von Bertalanffy, *A Systems View of Man* (Boulder: Westview, 1981), p. 111.

4. Milan Zeleny and Norbert A. Pierre, "Simulation of Self-Renewing Systems," in *Evolution and Consciousness*, ed. Erich Jantsch and Conrad H. Waddington (Reading, Mass.: Addison-Wesley, 1976), p. 150.

5. Robert Wright, *Three Scientists and Their Gods* [Edward Fredkin, Edward O. Wilson, and Kenneth Boulding] (New York: Times Books/Random House, 1988), see esp. pp. 67-69.

6. William F. Allman, *Apprentices of Wonder: Inside the Neural Network Revolution* (New York: Bantam, 1989).

7. Wright, pp. 79, 64.

8. James Jeans, *The Mysterious Universe* (New York: Macmillan, 1948), p. 186.

9. Rainer Maria Rilke, "Letter to Witold Hulewicz, Nov. 13, 1925," in *The Selected Poetry of Rainer Maria Rilke*, ed. Stephen Mitchell (Vintage, 1984).

10. Joseph Campbell, *The Inner Reaches of Outer Space* (New York: Harper & Row, 1986), p. 19.

11. Merrell-Wolff, p. 56.

12. Joseph Campbell, *Creative Mythology*, vol. 4 of *The Masks of God* (New York: Penguin, 1968), pp. 259-60.

13. G.W.F. Hegel, *Early Theological Writings* (Chicago: University of Chicago Press, 1948).

14. Quoted without reference in Eknath Easwaran, *The Compassionate Universe* (Petaluma, Cal.: Nilgiri, 1989), p. 130.

15. C. S. Holling, "Resilience and Stability of Ecosystems," in *Evolution and Consciousness*, ed. Erich Jantsch and Conrad H. Waddington (Reading Mass.: Addison-Wesley, 1976), p. 91.

16. See Daniel B. Botkin, *Discordant Harmonies: A New Ecology for the Twenty-First Century* (New York: Oxford, 1990).

17. See Beatrice Bruteau, "From Dominus to Amicus," *Cross Currents* (Fall 1981).

18. Matthew Fox, *Original Blessing* (Santa Fe: Bear, 1983), pp. 183-84, citing 2 Corinthians 3:18.

7

A Loaves and Fishes View of Productivity

RICHARD C. HAAS

The Environment as an Economic Issue

Too often discussions about environmental problems are structured in terms of jobs versus snail darters; that is, the practical economic interests of workers and their families versus elitist concern for inconsequential plants and animals. In this dichotomy economic development and environmental protection are treated as antithetical concepts. As a result, crafting environmentally sound economic policies and economically sound environmental policies becomes very difficult.

This polarized thinking about economics and the environment poses a special challenge for Catholics intent on promoting a vision of harmony between humanity and all other creatures. In order for us to have an impact on society's ecological awareness, we must be prepared to counter the underlying economic rationale which prevents proper development of all natural resources—mineral, vegetable, animal, and human. The economics used to justify environmental destruction regards wealth as a commodity to be extracted from the earth and from people. Fortunately, the Catholic tradition contains a powerful antidote to this limited and ultimately destructive view of material wealth. We belong to the loaves and fishes school of economics. Our core beliefs, the Trinity and the incarnation, tell us that love cannot be contained. Love is overflowing. It is abundant. The supply of love is inexhaustible. It is infinite. It is God. Thus, those of us approaching the environment and the economy from a Catholic perspective are in a unique position to draw attention to the multiplying effects of loving relationships.

Although talking about the multiplication of resources appears to be the ultimate exercise in "voodoo economics," building a sound understanding of the economy around the parable of the loaves and the fishes is not as preposterous as it might seem. When we focus on the most critical question about the economy—how can we make it *grow*?—we start to recognize that we are dealing with a living organism. Whereas economics is preoccupied with the too little and too few (income, jobs, resources), the life sciences often are confronted with too much. Except for the tendency of living cells to propagate, AIDS, cancer, and overpopulation would not be problems. Since the economy is a collection of living people, why are the notions of generation and multiplication, which are central to the study of biology, not applicable to the study of economics as well?

By emphasizing the reality and importance of human fruitfulness, Catholics have an opportunity to foster organic, natural, pro-life thinking about the expansion of wealth. Reflecting on productivity, the most fundamental of all economic concepts, in light of such fundamental Catholic beliefs as creation, the Trinity, and the eucharist, offers a way of comprehending wealth formation as a generative rather than an extractive process. The new understanding of productivity that flows from such reflection provides a conceptual framework that resolves the tension between economic development and environmental protection. A more refined notion of productivity also suggests new arguments for key elements in the Catholic economic justice tradition, particularly the living wage and the preferential option for the poor. Moreover, the focus on productivity can lead Catholics to a new appreciation for humanity's economic instincts. Recognizing the self-denial involved in voluntary productivity can help us to see the longing for true prosperity as another sign of grace drawing people into community.

In keeping with the Trinitarian imagery that is so important to the view of productivity that will be presented, the reflection will be structured around three arguments:

1. Productivity is about creativity not efficiency.
2. Productivity can best be understood within personal relationships. Particular attention will be paid to the Trinity and to sexuality as relationship models which can illumine our thinking about economic interactions.
3. Productivity cannot be mandated; it must be inspired by management practices, government policies, and a distinctively Catholic spirituality.

Productivity as Creativity

That religious thinkers have left reflection on productivity to economists is unfortunate, since this is a subject much broader than a simple mathematical measurement of output per worker. Productivity is a topic filled with wonder, for it touches upon the most mysterious of all human capacities, the ability of men and women to transcend their physical limits. Even the economist studying productivity must grapple with what is essentially a moral question: what prompts people to extend themselves for others?

Rather than introducing a new level of mumbo jumbo into economic discussions, a religious perspective offers a way of eliminating the confusion and contradictions evident in current business thinking about productivity.

For years, the need for improved productivity was one of the few points on which economists of all persuasions could agree. From the early 1970s through the mid 1980s, output per worker in the United States (adjusted for inflation) failed to increase despite higher education levels and technological improvements. Then, in a trend that accelerated during the 1989-1992 recession, businesses discovered new ways to get more done with fewer workers. By 1993, the U.S. economy was emerging from its most extended period of decline since the Great Depression, but the recovery displayed an unusual characteristic. Unemployment remained at historically high levels despite the quickening pace of economic activity. Even as the economy expanded, major companies such as IBM, Johnson & Johnson, General Electric, General Motors, and Delta Airlines— names once synonymous with employment security—accelerated their work force reductions. Where once business was criticized for not pursuing productivity with sufficient zeal, now critics are warning that the relentless focus on productivity could lead to a permanent downsizing of the American work force.

Productivity, Good or Bad?

So long as productivity is viewed as a technical concept measured by output per worker, contradictory answers to the above question can be supported. On the one hand, improved productivity offers workers their only chance to enjoy non-inflationary income growth. If they produce more, they will get paid more. As far as worker salaries are concerned, the higher the productivity level the better.

On the other hand, as output per worker increases, fewer employees are needed to produce the same level of goods and services. For this reason, organized labor has viewed productivity improvement efforts with great skepticism. Indeed, featherbedding is a deliberate attempt to lower productivity so that high levels of employment can be maintained. So long as one side advocates productivity improvements as a way to reduce employment, the other side will resist them for the same reason.

In one way or another all arguments over the best approach to "get the economy moving again" are variations of the productivity debate. Until a way to link higher productivity to higher employment is discovered, the American people will be presented with two conflicting visions of the future: a high-tech economy offering a limited number of high-paying jobs, or a low-tech, low-wage economy that offers broad employment.

Companies are right to recognize that their profitability depends upon the productivity of their workers. Yet managers who rely on the conventional output-per-worker measurement will always be tempted to see the elimination of all workers as their ultimate goal. Indeed, if the formula commonly used to monitor productivity is valid, the attitude that management's goal is to minimize employment is correct. In the equation (P)roductivity = (O)utput ÷ (W)orkers, the value of P approaches infinity as the value of W approaches zero.

The proper measurement of productivity, however, is not output, but value. Mathematically speaking, productivity is not a quotient but a remainder. It is much more closely aligned to such economic concepts as wealth and profitability than to efficiency. Productivity is the spread—the difference, the gap—between the value of an item (a good or a service) and the cost of producing that item. Even under the most precise economic definition, to be productive is to be creative, for productivity exists only when the value of output exceeds the cost of input. By definition, the productive person transforms the lesser into the greater. Productive men and women add value to the economy which did not exist prior to their efforts.

Unfortunately, our society's refusal to accept the miraculous—to acknowledge the human capacity for something-from-nothing creativity—blocks the application of productivity's proper definition. As a result, meaningful discussion about real economic development is very difficult. How does one articulate a prescription for true economic growth without accepting the *creation* of new wealth in the strictest sense of the word? Furthermore, the uncreative view of productivity is reinforced by analytical limitations. Determining the precise net value employees contribute to an organization is much more difficult than calculating the number of cars assembled or insurance claims processed. Thus, the shorthand figure of output per worker remains the accepted productivity benchmark, and the fallacy of trying to maximize productivity by minimizing employment continues.

Disputing those who underestimate men and women's economic powers and reversing the misunderstandings which glorify the workerless economy are among the most significant contributions Catholic thinkers can make to both economic and environmental understanding. As Michael Novak has noted, one of the most important strengths we bring to economic discussions is our strong belief in creation. We see ourselves both as creatures of an all-loving God and as co-creators of his kingdom. In *The Catholic Ethic and the Spirit of Capitalism*, Novak presents Pope John Paul II's 1991 encyclical *Centesimus Annus* as the culmination of Catholic thought on the importance of creativity in economic life. For the pope, the acting person—the creative person—is the decisive factor in the economy.

This is the type of thinking that can lead to a more complete understanding of productivity. Only when the reality of creativity is accepted does the productive worker (the one who generates value in excess of cost) receive the recognition he or she deserves. Such a person is quite literally an extruder of wealth, the real-life equivalent of the goose who lays a golden egg. Since each productive worker adds value to the economy, a society is not maximizing its prosperity so long as one potentially productive person remains unemployed.

Productivity in Relationships

Ironically, only in an economy that purports to be free does seeing workers' ability to generate value in excess of cost as the real meaning of productivity (and, by extension, the true source of all new wealth) lead to intellectual problems. In a controlled or slave economy, the value-added notion of productivity is hardly troublesome. Southern plantation owners went to war to defend slavery because theirs was a very productive economic system that contributed greatly to their families' prosperity. The demand for slaves was strong because there was a considerable spread between the value generated by the typical slave and the cost of owning a slave.

Karl Marx, for one, because he saw the economies of his time as another form of slavery, would have no problem with the definition of productivity as uncompensated human effort. Although many economists refute the technical points of his labor theory of value, Marx's underlying insight about the link between profits and underpaid workers is more difficult to ignore. Staffing a factory with robots represents a productive investment only if the people who build the robots are productive—that is, willing to pass on value in excess of cost. If productivity were measured in terms of net value rather than gross output, we would quickly discover that all productivity increases attributed to technology can be traced to the diligence of people who design, manufacture, install, and operate the new machines.

Some social justice advocates might argue that the above analysis of productivity verifies their conviction that an economy which finances its future growth with past profits is inherently exploitative. What could prompt a person to generate value beyond compensation other than some form of coercion? Certainly explaining this type of productivity in the traditional language of economic self-interest is difficult, if not impossible. Where is the self-interest in accepting less for more?

Yet to stop the discussion at this point would shortchange a fully Catholic understanding of wealth formation. The more difficult task is to explain workers' productivity—the gap between value proffered and compensation received—as their freely rendered contribution to the economy. And the hardest challenge is to shed light on that most mysterious of all economic phenomena, the mutually profitable exchange in which both parties simultaneously give more than they get and get more than they give. Tackling these intellectual problems, however, is necessary if Catholics are to offer a program for economic growth that maximizes respect for both persons and the environment.

As Catherine Mowry LaCugna noted in *God for Us: The Trinity and Catholic Life*, we Catholics rarely refer to the central mystery of our faith when trying to unravel the mysteries of life. This is certainly true of our probing of mysteries in the economy. Yet the Trinity provides an economic model that offers numerous insights into the expansive power of personal relationships. Through his life, death, and resurrection, Jesus Christ revealed to us not only his Father and their Spirit but also the inner dynamic of the Trinity. The relationship among Father, Son, and Holy Spirit is not one of static equality; instead, the Trinity pulsates with the rhythm of surrender and exaltation.

As St. Paul tells us in the Epistle to the Philippians, Christ emptied himself. He abandoned all claim to parity, forsaking the rights of Creator to become a creature. Yet this emptying did not result in the annihilation of Christ's divinity—just the opposite!

> Because of this,
> God highly exalted him
> and bestowed on him the name
> above every name,
> so that at Jesus' name
> every knee must bend
> in the heavens, on the earth,
> and under the earth,

and every tongue proclaim
to the glory of God the Father: JESUS CHRIST IS
   LORD! (Phil 2:9-11)

Thus, the Trinity provides us with a compelling lesson on love's explosive force. Creation took place because the Son

. . . emptied himself
and took the form of a slave,
being born in the likeness of men. (Phil 2:7)

The created universe exists because Jesus "did not deem equality with God something to be grasped at" (Phil 2:6). The divine willingness of the Son to transcend his self-interest makes the whole thing possible. Unless we believe Jesus' words that "whoever has seen me has seen the Father" (Jn 14:8)—that is, we believe that Jesus is both true God and true man—we cannot appreciate the depth of God's love for us and the heights to which we are called.

Belief in the Trinity helps us understand that behavior within loving relationships conforms to the backward logic of the cross, which redefines the meaning of self-interest. To die is to live, to lose is to find, to surrender is to be free. The power of personal relationships, therefore, is the key to a Catholic understanding of economic expansion. Alone, each person is confronted with personal limitations and material shortages. Yet within relationships—within community—the expansive energy of love is released. Offered to a succession of other people, finite matter acquires infinite usefulness. Passed through the right hands, five barley loaves and a couple of dried fish are enough to satisfy the multitudes. Harmony with other persons is the key to harmony with the material order.

If we are all independent economic agents and if each transaction must stand on its own economic merits, productivity makes no sense. There is no payback for generating value in excess of cost. Yet in the context of a relationship, such generous behavior becomes much more understandable. I will be far less concerned about coming out ahead on every deal if my relationship with my economic partners is marked by trust. Thus, expanding prosperity is largely a matter of fostering productive relationships.

Related to the Trinity is another Catholic image that suggests the qualities that make all personal relationships, including our economic ones, fruitful and productive. In Catholic theology the human relationship most closely resembling the divine relationship in the Trinity is the sexual union between husband and wife. How many of our views about the economy would change if we recognized that sex is not something extraordinary, but rather the most intense expression of a fundamental human reality: all personal relationships are meant to be mutually satisfying, mutually fulfilling, and mutually expansive?

Drawing a parallel between sexual and economic relationships provides a way of explaining how a relationship can be productive for both partners—how both partners can receive value that exceeds cost. The key is distinctiveness. In sexual intercourse the sensations of giving and receiving pleasure are indistinguishable. Asking whether the husband or the wife enjoys sex more is an irrelevant question, because each offers a sexual identity that complements the other. As a result, the pleasure of both spouses is mutual.

Sexual activity provides an example of the immense joy that can be ours when we respect another person as an equal partner. The radical difference of our spouse is the source of our sexual satisfaction. Yet sustaining awareness of this point requires considerable effort. In the midst of passion, accepting the otherness of our partner is easy. Over time, however, affirming our partner's equality requires abandonment of our false sense of superiority. Only an element of surrender allows the couple to preserve a long-term sexual relationship. Through their mutual abandonment, husband and wife achieve the Trinitarian experience, a sense of unity that enhances rather than diminishes each partner's individuality.

The Catholic church has been criticized for being preoccupied with sexual morality. Yet when sex is recognized as the archetype of all human relationships, the attention it receives does not seem excessive. The technical definitions of fruitfulness and mutuality may be open to theological debate, but the church's insistence that these characteristics mark every sexual act stems from an unassailable aim to preserve sex as an expression of giving, not of taking. One of the most serious problems with casual sex is that it confuses reciprocal abuse with mutuality in a way that infects other personal relationships. Many consumers, for instance, are susceptible to fraud because they operate according to the you-can-screw-me-if-I-can-screw-you philosophy.

In contrast, the productive economic relationship follows the self-surrender model of loving sex. Like mutual pleasure, the idea of mutual profit is grounded on the distinctiveness of every person. When viewed against the backdrop of human diversity manifested by gender, racial, and language variations, economic transactions become much more intelligible. If all people are the same, then one person must win and another must lose every time a buyer and a seller make an exchange. On the other hand, when we appreciate that each person possesses a unique set of gifts and a unique set of needs, we can see usefulness and value as relative concepts. Thus, the value a buyer can derive from a product rather than the cost of producing the product is the key determinant of price. I will be glad to buy from you so long as my evaluation of your wares is higher than your price. You should be eager to sell to me so long as my payment is higher than your cost of production. When these conditions exist, the relationship is mutually productive—value exceeds cost on both sides of the transaction. For these conditions to exist you and I must know each other's needs well enough so that we can create value for the other person in a way that does not exceed our respective costs. As John Paul II noted in *Centesimus Annus*, the "ability to perceive the needs of others and to satisfy them" (no. 32) is among the most important skills needed for participation in the economy.

Achieving an understanding of productivity predicated on respect for the distinctiveness of another person allows us to respond to the uneasiness with which many religious people regard economic life. What are we to make of the fact, they ask, that all economic activity appears to be "seeking more" behavior? Is not the fundamental law of the marketplace to trade up? How can justice prevail in the economy so long as every person is trying to get ahead?

Actually, the demand for more that permeates the economy is the basic condition requiring productivity. The market does not clear when value equals price, but only when value is greater than price. In a truly free economy, only companies which give their customers more than they pay for can stay in business, and only workers whose efforts exceed their wages will have employment. Yet the opposite conditions also will prevail in a truly free economy. Only buyers who pay more than cost will be able to make purchases, and only employers who overcompensate will be able to attract workers.

Within a context of freedom and respect for the other person, the drive toward economic advancement becomes a powerful stimulus for economic generosity expressed as productivity. Because fruitfulness is a condition of relationships, not individuals, the longing for prosperity as a sense of fullness and harmony with the material order becomes a powerful magnet attracting people together. The economic instinct, no less than the sexual instinct, draws persons into community. In economic relationships, as in sexual relationships, the surrender associated with accepting the other person as equal is the key to sustained satisfaction.

When freedom and respect are no longer present in relationships, however, the energies which draw people together are distorted into forces of conflict and destruction. Without the critical element of surrender, the sexual urge degenerates into lust and the economic urge degenerates into greed. Both of these sins are manifestations of our attempt to gain what the other person has without sharing of ourselves.

Fortunately, the Catholic community is blessed with the witness of evangelical poverty, chastity, and obedience, which cautions us about the way sin distorts our natural inclination toward fulfillment. Dedicated abstainers remind the sexually and economically active members of the community that the object of every person's deepest desire is neither money nor sex nor personal freedom. Vowed religious help us see our sexual and economic urges as expressions of a yearning for fullness that can only be satisfied through community with other persons, both human and divine. They also help us recognize that a prerequisite for accepting the contribution others can make to our satisfaction and fulfillment is self-denial— abandoning all pretense to sovereignty and superiority.

For those not called to lives of poverty, chastity, and obedience, the witness of the vows is a call to activism, not abstinence. As the Catholic theology of marriage has moved beyond sexual minimalism, we have come to see the important connection between ecstasy and self-transcendence. We know that embracing another person provides a unique opportunity to embrace our creaturehood in a truly salvific way. Now, the challenge is to develop a parallel theology of economic passion that accentuates the importance of a joyful, satisfying relationship with the material world. When we recognize that our innate drive toward prosperity propels us into productive (that is, self-surrendering) relationships, we see that our powerful economic urges—no less than our powerful sexual urges—are indications of a grace-filled universe drawing us into community.

Inspiring Productivity

One of the drawbacks to using sexual imagery to explain productivity is that it invites the observation that neither love nor respect nor mutuality need be present in the act of conception. The most fundamental act of creation can be forced upon another person in violation of her freedom. Thus, even accepting human productivity as the way in which new wealth is generated does not necessarily lead to the formation of true economic communities. As noted earlier, slave economies have been and can be productive.

Yet the comparison between sexual and economic relationships provides us with a way to evaluate measures aimed at raising worker productivity. Relying on fear, force, and deceit to increase the value of workers' economic contributions makes as much sense as recommending rape and seduction as the best way for society to replenish its population. Making love might not be the only way to make a baby, but the esteem with which all cultures regard the institution of marriage suggests that stable, loving relationships are the best producers of a society's future generations. Similar relationships are the best producers of a society's prosperity.

Understanding the economic as well as biological dimension of human fecundity is only the first step in charting a course toward prosperity. The hard part is developing the corporate, government, and religious practices that will foster productive behavior. Vatican II's *Constitution on the Church in the Modern World* offers a schematic on married love which can be applied to productivity. The chapter on "Fostering the Nobility of Marriage and the Family" looks at the union of man and wife from three perspectives, which show how the sexual relationship between a man and a woman evolved from a natural impulse to a social institution and finally into a sacrament. The primary experience, the council fathers remind us, is the "conjugal covenant." Through the institution of marriage, society protects this covenant so that it can enhance the "dignity, stability, peace, and prosperity of the family itself and of human society as a whole." Finally, "Christ . . . the Savior of men and the Spouse of the Church comes into the lives of married Christians through the sacrament of Matrimony." Through the sacrament "authentic married love is caught up into divine love."

Just as the magnetism of a man's and woman's sexual differences attracts them into a single conjugal covenant, the multiplicity of our material needs draws us into an unlimited number of economic covenants. In order for prosperity to flourish the institutional protection of society and the sanctifying assistance of the church that has been offered to sexual relationships must be extended to these economic relationships as well.

Inspiring Productivity—Management Practices

As noted earlier, the operative word for discussing economic activity is *more*. In seeking ways to encourage the productivity that will lead to a less destructive and more prosperous society, we must abandon our traditional notions of balance and even fairness so that we can make room in our discussions for the unbalanced idea of generosity. Traditional descriptions of the relationship between workers and employers, such as "full day's work for full day's pay" and "giving employees their just due," are inadequate to explain the dynamics of the loaves and fishes economy. What is needed is *more*. In asking workers to be more productive, companies are asking them to be generous—to give more than a full day's work for a full day's pay. Such generosity, however, cannot be mandated—it must be inspired. Threats and coercion are not the tools which will maximize prosperity. If companies want their workers to be generous, they must be prepared to reciprocate. They, too, must offer more.

An example of such generosity can be found in the parable of the laborers in the vineyard, which contains Jesus' description of a very unconventional compensation arrangement. Even the last hired, who gave far less than a full day's work, receive a full day's pay. Many of the workers regard the owner of the vineyard's action as grossly unfair. After all, they worked "a full day in the scorching heat" (Mt 20:12). Yet the owner responds, "I do you no injustice . . . are you envious because I am generous?" (Mt 20:13-15). No one said radical changes in personnel policy would be easy.

The living wage doctrine, which has been such an important part of Catholic social teaching in the past century and which flows directly from this parable, offers an extremely effective mechanism for motivating economic generosity. The productive workers so desirable to companies are people who do not count the cost at every occasion—employees who will take on extra assignments and expend extra efforts to meet important goals. This attitude of willingness and cooperation, however, can only be sustained within a climate of security and trust. In her wisdom, the church is offering employers an excellent piece of management advice. Pay people enough so that they do not have to worry about feeding, housing, and educating their families, and they won't quibble about their work. Once people have a sense of economic security and sufficiency in their lives, they are far less inclined to question whether they are adequately compensated for each specific assignment. Only then can employees afford to be generous in their work.

In contrast to the "lean and mean" tactics many companies are using to extract greater productivity from their workers, the living wage approach looks hopelessly idealistic. Yet the we-won't-count-the-cost-if-you-don't relationship embodied in the living wage doctrine is the exact type of arrangement successful companies try to structure for their most prized employees. In *The Firm* John Grisham describes the way a corrupt law firm, Bendini, Lambert & Locke, recruited the book's hero, Mitchell McDeere. The senior partners of the firm know that the best way to command an employee's unquestioning loyalty is to give the person *more* than he or she expects. Bendini, Lambert & Locke was not content to match Mitch's best offer—the firm topped it by 10 percent and then threw in a black BMW and a low-interest housing loan for extra measure. Like the fictional Bendini, Lambert & Locke, real-life companies offering their senior executives extremely generous compensation packages justify these arrangements by saying they don't want their key employees worrying about making money for themselves; they want their best employees to concentrate on making money for the firm.

As the gospel suggests, the wisdom of such wily managers should not be ignored. If companies are confident that a million-dollar employee will generate value far in excess of cost, why do they doubt that a much more modest living wage for other employees will produce similar returns? One of the major ironies of business life is that the best-paid employees in many organizations are actually the least expensive, because they are the most productive; the lowest-paid workers are often the most expensive, because they generate minimal value beyond their compensation. This generalization is used to widen the salary gap between those on the top-rung and those on the bottom-rung of the corporate ladder. Yet companies that follow this approach fail to see generous compensation as an inducement to productivity—not just a reward for it.

What is emerging at some companies after their downsizing is a two-tiered employment structure: an inner cadre of secure, well-paid, highly motivated managers, and an outer group of employees who are viewed as expendable. The productivity of the former group is taken for granted; they get a great deal from the organization but give back even more. The productivity of the latter group, however, is subjected to intense scrutiny. The expendables live with the constant fear that as soon as a worker fails the latest efficiency test, he or she will be replaced by a machine or a "temp."

Even managers who forsake the use of blunt terror tactics to prod their employees often miss the real point about low productivity by relying exclusively on training and technology as the solution to their problems. Productive employees are not necessarily equipped with more smarts or better tools than their unproductive counterparts. They are productive primarily because they are willing to give more. They have learned the one essential lesson of a prosperous society: on balance, participants must put back into the economy more than they take out of it. Prized employees exhibit the same willingness to surrender that we find in a loving marriage. They stop counting the cost and place their interests in the hands of other people.

Yet surrendering one's interests without the expectation of reciprocity is not an invitation to love—it is an invitation to exploitation. The church's proscription against sex outside of a permanent marriage represents the height of consumer protection. Surrendering one's sexual identity for anything less than a permanent commitment is a bad deal. This link between surrendering and commitment holds in the economy as well. Too few managers who complain about the absenteeism and inattention of minimum-wage workers recognize that all employees have the same economic needs. If they did accept the similarity between their workers and themselves, managers would see that extending tier-one treatment to the rest of the labor force would boost productivity throughout their companies. The living wage paid to every employee puts all workers in the position where they can concentrate on generating additional economic value.

Even managers who question the wisdom of using generous compensation packages as a way of inspiring both blue-collar and white-collar employees to greater productivity should appreciate the fact that the link between pay and performance is a complex rather than simple relationship. On a graph it would be represented by a curve rather than a straight line. Suppose we found that this relationship could be expressed mathematically as $P = C^2$, where "P" is productivity and "C" compensation. In this case we would discover that productivity is always less than compensation so long as value for compensation is less than one. For instance, the value of $.9^2$ is .81, which is less than .9. On the graph the curve would be almost flat, since each increment of productivity would be less than each increment in compensation. Yet once compensation went above one, productivity would jump exponentially ($2^2 = 4$, $3^2 = 9$, and so forth). The curve would start shooting almost straight up.

Unfortunately, there is no magic formula that will help companies optimize the productivity of their employees. Yet the above example illustrates a point that business people involved in management training can verify. The transforming point in the career of a would-be manager often comes with a sudden click. At that moment the person establishes a bond with the organization and starts to identify the employer's interests with his or her own interests. Inspiring that click is what improving productivity is all about. Until a company's total compensation package (including working conditions and personal respect as well as wages) reaches "one"—the all important unit value—the response of "ungrateful" employees may be discouraging. Yet when a company inspires an employee to cross that mysterious threshold of loyalty, a personal relationship is established. Then each expression of the company's generosity is compounded by the employee and returned in the form of higher performance. From that point on (or until the employee's trust is violated) the veritable magic of productivity occurs. Value exceeds cost by a wide margin.

What the church is suggesting through the living wage doctrine is that the ungenerous company is wasting its money. Only organizations willing to establish a personal relationship with employees by providing them a generous measure of economic sufficiency will be able to inspire true productivity. Downsizing and wage cuts are not the answer to the productivity problem.

Inspiring Productivity—Government Policies

Business cycles in the economy of the loaves and fishes conform to the rhythm of the Paschal Mystery. Defeat is as much a part of our economic lives as success. In light of this principle we must recognize that even companies that treat their employees generously and respectfully may not be able to offer the security of a lifetime contract. In an organic economy, dead branches are pruned so that new shoots can sprout.

Thus, establishing a society's productivity-inducing atmosphere is a responsibility that supersedes the ability of individual corporations. If productivity and prosperity are to be abundant, society as a whole must work to establish a pervasive climate of goodwill. In order for people to become productive, they must have the sense that the odds favor the giver—not the taker. Viewed as inducements to productivity, law and justice, which receive such strong emphasis in Catholic social teaching, take on new economic importance. Although voluntary productivity is radically difficult because it involves surrender and self-denial, the moral society has an obligation to facilitate this form of economic generosity.

One important contributor to this sense of goodwill is a strict standard of profitability, which focuses society's attention on the formation of real wealth and rewards productivity. Such a standard would eliminate much of the confusion about spurious economic activity. Crime, fraud, corruption, and environmental destruction are tolerated in many parts of our society because these activities are associated with economic benefits. Drug dealing provides jobs in the inner city, insider trading expands incomes on Wall Street, strip mining boosts the economy of Appalachia, clear cutting benefits loggers in the Northwest.

A broad system of accounting is needed to distinguish those endeavors which make a net contribution to society from those which do not. The challenge of business is to operate profitably and thereby create new wealth. Yet no business has an innate right to make a profit. The net contribution of a company that cannot cover its full environmental and social costs is negative. Therefore, it has no economic value. Unfortunately, an economic system which makes job creation rather than wealth creation its primary objective exposes society to extortion.

Only in the light of a clear understanding of profits can the economic damage of crime and pollution be illuminated. All endeavors that inflict costs on society in excess of the benefits rendered are forms of theft which block the drive toward prosperity. Through consistent law enforcement, government informs criminals, con artists, and polluters that they have contributed nothing to society and thus deserve nothing in return. Yet when laws are not enforced and the extractors are allowed to benefit from their rape and deceit, the attractiveness of productivity is diminished and the likelihood of people extending themselves for their economic partners is reduced.

The ultimate expression of a society's benevolence is its treatment of the unproductive. By giving to those who cannot be expected to give in return, society establishes the generous atmosphere that encourages its members to become productive. The very fact that they have nothing to give makes the poor valuable to the loaves and fishes economy. The poor create an opportunity for society to exercise true charity—to give without the expectation of return. Once assisted, the poor become the most important communicators of a society's goodwill, without which a society cannot expect to foster productivity and prosperity.

In this context, the church's concept of a preferential option for the poor takes on economic as well as moral significance. The benevolent society intent on productivity is willing to assume obligations where no obligations seem to exist. Such a society acknowledges the rights of the poor described by the U.S. bishops in *Economic Justice for All*: "life, food, clothing, shelter, rest, medical care, and basic education."

At the time the pastoral letter was first drafted in the mid 1980s, the bishops' notion of economic rights met with strong resistance, and the prospect that these rights would be established by law seemed remote. Yet, now a consensus is emerging that the right to medical care, potentially the most expensive of the bishops' recommended rights, does indeed exist. The majority of business leaders who already offer their employees health benefits are now among the strongest advocates of health-care reform. They recognize that a system of universal health coverage would introduce a positive measure of fairness and flexibility into the American economy.

This chapter is being completed a few days after the formal introduction of President Bill Clinton's health-care security plan. Thus, the eventual congressional response to his proposal is still unknown. If adopted, the program will provide a fitting test for this presentation on the interplay between goodwill and productivity. By addressing workers' fears about their long-term ability to afford quality medical care, the program could have a dramatic impact on workers' willingness to take risks and expend effort. If raising workers' sense of health-care security has the predicted effect on their productivity, the concerns about the cost of health-care reform may prove to be unfounded. Should the commitment to universal health care turn out to be a productive investment for our society—that is, the benefits exceed the cost—the nation will have received an extremely important lesson in the economic value of generosity.

Inspiring Productivity—Catholic Spirituality

Although there are important steps both corporations and government can take to inspire productivity, the generation of value in excess of cost is an action which can only be performed by the individual human person. The management and politics of productivity must be reinforced by a spirituality of productivity if this type of behavior is to become widespread. Nothing short of a total rethinking of our economic behavior will allow us to practice this productivity consciously and consistently. The economy, now regarded as a place where we gain, must become a place where we give.

For Catholics seeking to develop an integrated spirituality that energizes all of their relationships—with the Creator as well as all creation—productivity can become an economic expression of holiness. When freely rendered, productivity is similar in many ways to Pope John Paul II's concept of solidarity. "In the light of faith, solidarity seeks to go beyond itself, to take on the specifically Christian dimensions of total gratuity, forgiveness and reconciliation," the pope writes in *Sollicitudo Rei Socialis*. "One's neighbor is then not only a human being with his or her own rights and a fundamental equality with everyone else, but becomes the living image of God the Father, redeemed by the blood of Jesus Christ and placed under the permanent action of the Holy Spirit" (no. 40).

An attitude of "total gratuity" is the key element in both productivity and solidarity. This seeking to go beyond oneself opens us to a specifically Christian dimension of reality where all of creation can be seen in its relation to the Trinity. Fortunately, the central ritual of the Catholic faith is directed specifically at fostering this requisite sense of gratuity. Participation in the eucharistic celebration of the Mass is, therefore, Catholics' most important economic activity.

The first step in achieving prosperity is becoming conscious of the riches we have to share with others. That only the rich get richer is the most basic of all economic truisms. The Mass is a celebration of richness, which stimulates our sense of gratuity. It is a reminder that the most important item we have to share with our fellow men and women is fellowship with Jesus Christ and, through him, fellowship with the Father and the Spirit. No matter what our economic or physical circumstances, we have something to share. We are, therefore, rich.

Once we have activated and reinforced a sense of fullness through our eucharistic expression of thanks and praise, we are forced to share—not out of any moralistic compulsion, but as an automatic, almost physical reaction. When we are full, we overflow to others.

The Mass celebrates and encourages sharing by focusing on the ultimate act of sharing. We express our faith that Christ not only accepted a share in our humanity, but also offered the promise of a share in his divinity. The eucharist reminds us that the second person of the Trinity has accepted our otherness, our apartness from God—our radical imperfection—and in exchange has offered us his sameness with God. In the Mass, our reaction and response to this God-with-man sharing is to enter into deeper communion with those around us. We share a meal, we share a sign of peace and fellowship, and we share our common bond with all people—past, present, and future—whose lives are marked with the sign of faith.

As important as this eucharistic attitude of gratuity is, it alone is not sufficient for full and effective participation in the economy. Like sexuality, economic activity is a reflection of life's pervading polarity. At the most basic level, all particles in the universe inter-act through forces similar to electromagnetism, which depends upon a combination of positive and negative charges to create its binding attraction. In identical fashion, unifying personal relationships require a double stimulus. Partnerships are sustained through a combination of giving and getting.

Gratuity energizes the positive pole of the economic development circuit. The eucharistic posture of thanks and praise provides the strong experience of having which must precede acts of giving. Yet deprived of a counterbalancing force, a sense of giftedness can deteriorate into smugness, which triggers economic isolation—not unifying interaction. An electric current cannot flow through a circuit with only one pole no matter how strong the positive charge. The negative pole necessary to complete the circuit and stimulate economic activity is an equally strong sense of neediness and dependency.

In addition to the Mass, which stimulates gratuity, the Catholic spiritual tradition is filled with imagery reinforcing humanity's sense of nothingness, its creaturehood. At the center of this tradition is the figure so important to Catholic spirituality—Mary. Her Magnificat outlines the inverted logic of God's salvific plan. Mary in her "lowliness" provides an opportunity for God to manifest his "greatness." The "proud" are "confused." The "mighty" are "deposed." The "rich" are "sent empty away." Yet the "lowly" are "raised...to high places," and the "hungry" are "given every good thing."

Mary, the true daughter of Israel, accepts her status as servant. God, however, does not treat her as a slave. "All ages to come will call me blessed," she tells us. "God who is mighty has done great things for me."

Mary's poverty creates the possibility of richness, because it opens her to God's grace. Her emptiness creates a vacuum attracting "all good things." The negative charge emitted by her lowliness attracts the positive response of God, whose "mercy is from age to age on those who fear him."

Through her canticle Mary expresses the reality that the worst thing that can be said of humanity also is the best: we are creatures, totally other than God. If God is all-perfect, we are all-imperfect. Yet, because each of us is a free individual, separate and distinct from God, we extend God's presence. Like Mary, we, by our very existence (independent of any behavior), are augmenters of God's glory. We magnify "the greatness of the Lord" (Lk 1:46-55).

The example of Mary and the other saints who find joy in their creature status has economic ramifications for both the poor and the wealthy. Poverty represents an essential side of the economic equation. Without a substantial volume of unmet needs, the market cannot exist. Receivers are as important as givers in the economic exchange. Thus, the experience of poverty can prompt a productive relationship when it is coupled with a spirit of openness to the mercy and love of other people. The emptiness of the poor provides a clear channel through which goodness can flow.

On the other hand, the rich will be rebuffed by the market—they will be sent away empty—unless they develop an awareness of needs to balance their manifest gifts. The economic participant who refuses to buy eventually runs out of partners to whom he or she can sell. Only the continuous matching of gifts to needs and needs to gifts can sustain a long-term economic relationship.

Both poles of the circuit must be kept clean from corrosion if the current of economic progress is to be sustained. The balance between an experience of fullness and an experience of emptiness motivates our economic interactions. Through the eucharist and devotion to Mary the Catholic tradition provides strong images of gratuity and creaturehood to nourish both aspects of economic spirituality.

Seeing productivity as an economic expression of holiness high-lights the generosity and self-surrender involved in this behavior. Yet a spirituality of productivity also confirms the possibility of attaining holiness through our everyday economic activities. The productive person is not an economic freak. We enjoy the level of prosperity that we do because the economy is filled with produc-tive people. They are the crafts people who constantly sharpen their skills as a way of improving the quality of their labor. They are the service personnel genuinely interested in being of assistance. They are the managers who take responsibility not only for their own personal performance but for the work of those they direct. They are the marketers whose primary focus is on the needs of their customers. They are the investors who risk their capital to finance worthwhile ventures.

Every participant in the economy who is willing to give without a guarantee of getting and who is motivated by the hope rather than the certainty of a reward is a productive person. The uncompen-sated efforts (the value beyond cost) of these productive people become the grains of wheat which fall to the ground and die so that prosperity can blossom. Achieving prosperity through productivity rather than plunder is not an unattainable ideal. It is an approach taken by many people in many parts of the globe. Yet clear think-ing about the human capacity for creativity and self-sacrifice is essential if productivity is to be adopted as the universal norm for economic behavior and worldwide prosperity is to be adopted as the universal goal.

At the heart of the Catholic faith is the belief that through the intervention of Jesus Christ in history and through the continuing presence of his Spirit, God shares with humanity his divine es-sence, the ability to love. By reflecting on the wealth creation process in light of this belief, Catholics can offer the world an economics that maximizes prosperity by maximizing respect for all of creation.

8

Appropriate Technology and Healing the Earth

ALBERT J. FRITSCH, S.J.

Revealing the current environmental crisis is dispiriting if unaccompanied by ways of overcoming it and beginning the process of healing our damaged earth. Messages of doom are depressing and paralyzing if not coupled with the good news of regeneration, restoration, and making this suffering planet better than it was before. By *earth healing* we mean meaningful repair of environmental damage, either through individual actions, such as stopping soil erosion or curtailing the use of chemical pesticides on a particular field, or through communal or collective actions, such as reclamation of surface-mined land or installation of industrial air pollution equipment. More important, healing needs to introduce and promote alternative human practices that use fewer resources and fit more into the ongoing natural processes of earth healing, namely those that are ecologically appropriate.

Both environmental pollution abatement procedures and alternative methods of regeneration require enthusiastic and hope-filled agents of change along with proper tools and methods; good intentions unaccompanied by systematic remedies may further damage a fragile environment and depress its inhabitants. We need to advocate the cause of environmentally suitable methods and appropriate technologies, which exist but are not nearly as well publicized or promoted as more sophisticated, costly, and complex modern devices and gimmicks. In part, this is due to a consumer culture that finds the more sophisticated, higher priced, expert-controlled and planned-obsolescent techniques to be more profitable.

Appropriate or people-centered technology became the concern of a Catholic economist, the late E. F. Schumacher. He was one of the first to see the great post-World War II flaw in introducing steel mills and tractors into lands which did not have the infrastructure to maintain and service them. He noted the tendency to discard highly suitable traditional methods in the mad global rush to modernity. He and others who worked in the Third World saw dignity and value in time-honored simple techniques being dismissed as old fashioned—even though they could do the job quite well. And he recognized the difficulty in reversing that trend.

A number of technological characteristics have been enumerated that will lead to genuine earth healing. These are found in technologies that

- take small amounts of capital;
- strive for locally available materials (lower cost and fewer supply problems);
- are relatively labor intensive;
- tend to be small enough in scale for families and small groups to afford;
- are understood, controlled, and maintained without a high level of special training;
- can be produced in small workshops;
- assume that people can and will work together to bring about community improvement;
- furnish opportunities for local operators to become involved in modification and innovation process;
- are adaptable to different places and changing circumstances; and
- are environmentally benign.[1]

Note that these are technologies of the poor, those to whom the church gives a preferential option, and those this author believes are most instrumental in the healing of the earth.

Without attempting to compare or negate the contributions that can be made to this discussion by other groups, religious philosophies, and institutions, I will attempt here to outline four strong Catholic actual or potential contributions to the grand enterprise of earth healing, which will be discussed in turn in the following four sections.

First, people steeped in liturgical practice find technology not something to fear but to champion, for when properly controlled it can be a form of liberation for all people by providing for the necessities of life with greater ease and consistency. Thus, liturgically trained people perceive technology neither as oppressive and to be avoided nor as an idol to be worshiped, but rather as a salutary means for healing the earth.

Second, our Catholic faith encourages us to seek truth actively through the study of nature as part of the good news, thus requiring our close cooperation with all people of good will. This is not some vague mandate, but one which stems from our evangelistic mission and which perceives healing the earth as involving all peoples, not just the elite. We profess a profound pro-life stance that naturally extends in concern and protection to all plants and animals, championing their rights to exist and flourish. This concern and protection require suitable tools and technologies.

Third, the church honors tradition by accepting the good of the past as a means to the future and as part of the ongoing process of salvation. Through prayerful discernment of appropriate from inappropriate technology the church promotes a balanced view of technology—neither a Luddite forceful return to primitive conditions nor an uncritical embrace of a throwaway mentality that champions the modern and relegates the old to history's dust bins. This tendency to avoid extreme positions is necessary for conserving older technological methods and allowing for ongoing improvement of the human condition. In cases of inappropriate technology the church is aware that healing only comes through dealing with our past rather than repressing it.

Fourth, church institutions exist all over the world and are able to network, furnish information, and exchange ideas, especially among third-world and lower-income people, who comprise the majority of its members. Through this attention to the poor and the long experience of global relief and charitable efforts, the emerging servant church realizes a commitment to mobilize resources for the actual earth healing process. The special Catholic flavor given in this work is one of excitement, celebration, and the confidence and joy of the risen Christ already present in our midst.

Technology as Eucharistic Response

From youth many believers have pondered the connection between liturgical celebration and application to daily life. It is always a mystery that grows on us but is never fully answered. It is no accident, says Nicholas Zernov, that a people steeped in the celebration of the eucharist developed a technology.[2] Matter was not something foreign but could be grasped in their hands and through God's grace could be transformed into Christ himself. Through the process of liturgical education believers overcame their fear of matter, found they could control and harness it in transforming ways—and do this for the good of others and for the free time needed to reflect and pray. In the eucharist the goodness of God's creation is acknowledged, praised, and thanked through the offering of gifts of bread and wine made by human hands—not natural gifts. A believing people are empowered to make holy or consecrate and partake in a transubstantial event. Jesus Christ is present. The transformed gifts are assimilated in communion for the betterment of all present, so that all participants may go forth and spread the good news, thus extending the people's experience, for *liturgy* means "the work of the people."

The salutary role of the monks in the Dark Ages, who brought stability out of chaos, has never been fully appreciated. The Benedictine tradition of *ora et labora* ("prayer and work") mingled divine and human creativity. These monks strived to build appropriate communities and firmly believed in the sacredness of tools and labor-saving devices. They changed the face of Europe using wind and water power. Following these technological innovations much adaptation was achieved in the lands they influenced. Thanks in great part to these monks, by 1086 C.E. there were 5,624 water mills in England alone, and virtually every community in Europe had one a century later. Northern Europe, with its seasonal use of water power due to winter freezing, became the haven for windmills introduced between 1150 and 1300 C.E. While classical Greeks and Romans recognized the power of wind, water, and even steam, they never made the connection of harnessing power to save human labor.[3] For the Greek and Roman elite class such ideas became playthings. For them human slave or free manual labor was demeaning, merely menial, of lesser value than their own cherished rational and supervisory activities. The Greek and Roman economic systems disdained laborers, even artisans, and never considered labor-saving devices of value, even though, as Casson says, labor upkeep and costs were high and laborers often scarce.[4]

Christianity changed all that, and it did so in part through a technology that the monks and nuns supported and taught through demonstration. The simple Christian rhythms of prayer and work extended beyond monasteries to the lives of all the people. They also were to be free to relax and celebrate communally even for relatively short periods and be able to break the routine of sunup to sundown labor. The church's championing of this right to pray in public led to establishing sacred times and places, and thus the strict demands of Sundays and feast days where no servile (slave or servant) work was allowed. A technology that saved labor (energy and time) became a tool or means to liberation, an integral part of the ideal Christian rhythm of prayer and work.[5] Technology made the coffee break possible. But the Medieval mind, nurtured by strong ecclesiastical and civil sanctions against servile work on Sundays and feast days, made prayer time an acknowledged real-

ity. Even amid imperfect performance this first stage of technological understanding was highly successful.

More important than even prescribed free time is the expectation that the believer of whatever class is to contemplate and pray
and that all are drawn to the deepest revealed mystery, namely the
Trinity. Thus all believers are marked with the sign of the Trinity
and repeat it often in prayer. In the following three ways we unveil
the Trinitarian mystery in our lives, and in all three we make use
of technologies at our disposal. We affirm our God creating and
fashioning the cosmos, redeeming and making good what is misdi-
rected, and enlivening and drawing out the fruits of each individual
believer. Without basic tools and conscientious agents there would
be no manifest human creativity; without appropriate method there
would be no healing or redeeming of the earth; and without communication systems there would be no normal conveying of
enthusiasm to others in near and distant places.

Laborers use technology in a creative manner to provide for necessities and amenities of life. Unfortunately, many of the modern
American elite have returned to thinking of labor in derogatory
terms; they consider the task of technology to enrich themselves at
the expense of others. The church needs to speak out and support
good technology and to restore labor to a place of dignity. Workers
can do this by proclaiming the Trinitarian mystery through creating, redeeming, and enlivening the environment in which they find
themselves. Laborers have been given special favor by God and
may fulfill in their work what is initiated in that liturgy. At their
baptism they are commissioned to praise the Trinity through their
hands as well as their lips. They are not mere bystanders in the
ongoing work of creation but actual doers who translate word into
action (see the Letter of James).

Creating. Through our call to heal our wounded earth we recreate even more wonderfully and prepare for the New Heaven and
New Earth. The surge of self-worth that homebuilders experience
completing their humble homesteads allows them to say, "We have
built this abode—through God's grace. We know what we can do
and will take the consequences for our own mistakes." Such persons experience growing self-esteem and are open to new
opportunities for creative growth. Through membership in a universal community of believers they may extend their homemaking
activity from the local level through a more collective *we* to a global
community. By first acting *and* thinking locally, they are availed
the opportunity to think and even act globally. Note the subtle
difference between this and "act locally and think globally."

Redeeming. Liturgy is a patient teacher. Those who participate
in the liturgy with regularity discover the rough spots in their spiri-

tual journey. They become aware of the need for redemption, of temptation, weakness, fall, the call to seek repentance and forgiveness; and the trust that God's forgiving hand will allow them to be better agents of change in the immediate and remote future. When our wrongdoing involves the resources of another, we need to make restitution of an individual or collective nature, depending on the manner of damage. Here we confront the social sin of ecological damage to God's good earth. Since liturgy celebrates the crucifixion and resurrection of Christ, we too must go down with him in order to rise with him. We are led to deeper levels of spiritual development in a pattern that follows the suffering and death of Jesus, and thus we are more able to take up the redeeming role of the Savior through reparation or restitutive activity.

Enlivening. Through prayer, song, music, and exhortation, liturgy brings about an uplifting of spirits that invites participants to become co-enliveners. The scarcity of genuine celebration is most evident in this tired world, which craves a spiritual uplifting. The media are saturated with bad news of earthquakes, wars, and murders. Communal uplifting and enlivening are not done through a single action but invite application of a variety of gifts which, in turn, require a variety of tools and techniques. Some earth healers are clowns and comics, others prophets, and still others Good Samaritans. Some become stewards of proper soil conservation or forest management, others educators with a vast array of research and teaching techniques, and still others regulators who must apply social controls to save precious environmental resources. The Spirit speaks in diverse ways, and no one is totally complete. An interaction of all the many psychological types of enliveners is needed so that the cooperative whole exceeds the sum of the individual parts and that interactive correctives be given to all participants in the earth healing process.

Technology as an Expression of Human Creativity

The divine ongoing creation story invites us to participate as co-creators making use of our unique talents and the tools that are available and discerned to be appropriate. The church shows concern about both the laborer and the tool used, and this concern for the human person and the means used has a long history, which extends to sacramental dispensation ánd preparation of ministers. Broadly speaking, the healing arts and the accompanying technologies dealing with ecological protection are similar to sacramentals that assist us in completing the healing task at hand. God works through us, and we are effective through properly chosen tools. As users, we need be in the right disposition, and that means humbly accepting responsibility for proper use and control of tools and methods and not allowing the technologies to control us. Through respectful use, technologies may be means to human liberation from the scourges of famine and disease. When improperly used, they *become* the scourge that denudes land, fouls rivers and air, and strips the earth of its valuable resources.

In 1967, just when believers were becoming quite proud of the Judeo-Christian contribution to modern technology, Lynn White, Jr., leveled a charge that our tradition was the root cause of the ecological crisis, due to its spirit of dominance and conquest of nature.[6] The Judeo-Christian activistic tendencies and propensity to discover, control, and explore had spilled over by some professed adherents into exploitation of the earth's resources, as labor- and resource-saving tools were transformed into technologies of profit, greed, and conquest over people and the creatures of the earth. When technological mastery finally arrived, we discovered the agenda had been coopted by the natural tendencies toward self interest among the undisciplined and immature. Though a number of writers have attempted to refute the dominance charge, this self-discovery has taken us beyond a second or imperial stage of our understanding of technology and to the dark night of the technological soul.

Did modern technology catch everyone napping, or more harshly speaking, does modern technology bedevil us with a secret power that entices users to a false sense of mastery? Have we conquered peoples and extracted resources that were subsequently wasted in frivolous uses? For those caught up in modern lifestyles, has the availability of low-cost electricity turning night to day blurred the daily rhythms? Have central space heating and cooling, along with transporting distantly-grown foods to the grocery market for supplies in any time of the year, dampened the impact of seasonal change? Have automobiles and other forms of transportation made this the most mobile of generations? Do we forget to ponder the ecological absurdities of the culture of waste and the throwaway container? And the final and most important question: Have these creature comforts and enticements dulled our creativity and our service role to the things of earth?

Once we accept that we have an ecological crisis partly of our own making, whether individually or collectively, we may respond by denying it, excusing ourselves as untrained or inexperienced, or escape from the problem through alcohol, consumerism, or forms of entertainment. Or we may confront the issue head on, responding almost instinctively with some form of "first aid," but not asking questions of how it happened or where the medics are. Poor suffering earth cries out, and we respond immediately because earth is our mother. Compassion is intertwined with the creative impulse. What will we use to heal the victim? Where should the victim be hospitalized so that healing can occur more readily? When will remedial action commence? Such questions suppose the existence of physical facilities, modes of transportation, medicine, and planning, in fact, the full scope of an appropriate healing technology.

One current tendency among critics of an activistic approach by those imbued with classical Greek and Roman cultural baggage is to concentrate on the many weaknesses of modern technology and to equate these with *all* of technology. From this classical bias these critics continue the perception that practical arts and human labor are beneath superior thinkers and to be relegated to the area of minimum wage earners. They realize, accurately, that modern technology has enhanced the macho mentality of speed, power, and fast getaway. Foresters openly admit that the chain saw has done more to threaten the world's forests than any previous invention. Powerful tools are powerful temptations. It is a rare and humble person who uses powerful instruments sparingly and properly. Since old-fashioned self-abnegation is seldom practiced, the fault continues to be borne by the dumb tool, not the unscrupulous operator.

A second current tendency is to regard modern technology as alien and hostile—beyond our reach. In fact, without generally admitting it, many are afraid of modern innovation as beyond their immediate control. The fearful may not have the dexterity to become a computer connoisseur, and they know it. For them, the instruction manuals are like a foreign language and the commands a threat to their balanced disposition. The more aggressive may buy into the culture all the more, and the more timid may surrender to having the expert do the job. However, none of us can excuse ourselves from responsibility.

Countering these tendencies while accepting the valid points raised may take a variety of paths—some more effective than others. Understanding the role of technology and learning to control it responsibly involve understanding our own place and time on this earth. Regaining a proper sense of place and time, the "here" and the "now" of our lives is a challenge. Our proper place is the immediate surroundings in which we live—the geological structure, watershed, climate, flora and fauna, migratory bird patterns, and local cultural history. Recovering our sense of direction requires focal points, of which our home serves as pivot and sacred space, a necessary place of rest and relaxation. The stability of "church" is important for those needing spatial ecological orientation.

Likewise, regaining our sense of season and time, the "now," demands a return to rhythmic patterns with recognized variation in the natural flows of the day and season. A return to liturgical worship and the stability of formal prayer patterns allow us to appreciate seasonal rhythms and the time of day and night for the rising and falling of our emotional lives. Sacred place and time afford us space and a temporal span for focusing on the presence of God. As we acknowledge our Creator in the goodness of creation, we become more conscious of our own unique place and precious time for caring for the earth with others. As the U.S. Catholic bishops wrote in the pastoral letter "Renewing the Earth," "By preserving natural environments, by protecting endangered species, by laboring to make human environments compatible with local ecology, by employing appropriate technology, and by carefully evaluating technological innovations as we adopt them, we exhibit respect for creation and reverence for the Creator."[7]

With this reemerging awareness of the sacred we regain consciousness of our place and time on earth and our relationship to all of creation. We are immersed in natural beauty, which inspires us to organize well the leisure time given us. In the rhythm of work and relaxation we come to appreciate God's grandeur in all the pristine forests, mountains, and rivers. If we ignore the beauty and stateliness of the oak tree or the warmth and music of the evergreen or the clean feel of a breeze on a mountainside, we will be unable to appreciate the contrast between human handiwork and human devilment, between enhancing nature and damaging it. Reflecting on God's creation opens us to enter into the creative process itself. On the other hand, busyness distracts or mesmerizes us and stifles our creativity. The rhythm of relaxation and work permits a pattern that makes us sensitive to our cultivated and desecrated environment and frees us to discover our call to action and healing.

Within an atmosphere of relaxation and reflection we discover a middle ground between disdain of technology and its uncritical use. This may be achieved through spiritual discerning, practicing self-control, and developing a healthy critique of what is modern. The unfocused and unthinking will extend environmental damage in fits of rage and further activity; the fearful will be paralyzed and intimidated by the rush of modern innovation. On the other hand, a healthy dose of activity and fear may help foster an ever-deepening respect for the Creator and the created and permit us to be more caring and protective of our fragile earth.

However, creation-centeredness by itself is not balanced: spiritual ecological balance demands creation *and* redemption; creation acknowledges blessing; redemption shows need to correct human wrongdoing. But it is not a mere juxtaposition of one with the other. Once accepting and entering into the sufferings of Jesus, the prime source of all revelation, believers are touched by his healing ministry—both the invitation to heal and be healed. This is regeneration, new life, resurrection. Earth is neither a powerful goddess, Gaia, nor a passive suffering being; rather, earth is the created and redeemed sub-stratum from which resurrection is occurring. "From the beginning till now the entire creation, as we know, has been groaning in one great act of giving birth" (Rom 8:22).

Technology as Establishing Eco-Justice

Environmental degradation and urban development have made it harder to find God's untouched creation. At the same time, experiencing untouched nature has become a prime value in our modern culture. An emerging business called eco-tourism directs people with time, money, and concern about the earth to venture forth and experience the vanishing pristine areas of the globe. They travel to uninhabited Antarctica, willing to tour and leave their cast-off products in a most fragile terrain. They vie to conquer remote and exotic geological formations such as Mount Everest, while the Nepalese government currently restricts the number of climbers because of environmental impact. A century ago such places were genuine hurdles, but this has changed because of the efficiency and speed of modern transportation. A burgeoning new enterprise brings thrill-seekers to the remote untouched, and through their touring makes pristine places all the more rare.

Some question the economics of eco-tourism almost as much as they question forms of resource exploitation. While a controlled modern tourist industry may minimize environmental impact, many forms of resource extraction do lasting damage. Let's reconsider the chain-saw operation. When loggers were equipped with axes and crosscut saws, selective logging was virtually their only option. Not so with the powerful chain saw, which topples trees and undergrowth at record speed and little effort. Clear-cutting the forest area for log removal was unthinkable in times past. However, an impatient chain-saw-wielding logger can sweep away decades of growth and even ruin the fragile ecosystem in record time. Such loggers fail to recognize that while trees are renewable, fragile forests essentially are not.

Also consider the use of agricultural chemical sprays. In Peru I asked why certain solar food cookers couldn't be introduced in the higher elevations where kerosene was scarce and solar energy plentiful. The response was that the people were too traditional to make such radical changes. I remarked that the ubiquitous pesticide spray backpacks had been widely adapted in only five years—but the guides countered that the multinational chemical company's advertisements were highly effective. Truly spoken. Agro-chemical addiction comes easily in diverse cultures, and reversing the trend is difficult. While over one-tenth of America's cultivated land is becoming no-till, the trend to reducing chemical herbicides has been reversed now and is on the increase again. Is chemical dependency one thing and solar adaptation another? The time is ripe for an examination of our social conscience.

It seems that we have a weakness for doing the wrong thing with the vast resources of our earth. Little wonder that the story of the Fall comes so closely after the creation story. Amid the wonder of creation splendor we need redemption. A collective examination of conscience reveals the wages of social misdeeds: polluted streams, acidified air, ozone-depleted atmosphere, denuded forests, eroded slopes, untreated and even expanding hazardous waste sites, scarred hills from surface mining, an increasing number of endangered and extinct species, and other threats too numerous to list. These are examples of what Edward Echlin calls the defiant act of ecological sin.[8]

It takes faith both to confess and to have the confidence that divine forgiveness is possible and occurs. Being shriven includes a willingness to make restitution for environmental damage both to current living human and other creatures and to unborn generations. Only with genuine forgiveness can eco-justice be reestablished on earth. To recreate and enter into the co-redemptive process involve our choice of good instruments, especially where the operation is quite delicate. Furthermore, we constantly remind ourselves that this healing process is a fruit of our eucharist, for it is a cosmic regeneration begun in the blood of the Redeemer.

Because of the ongoing need for social discernment, spirituality plays an important role in the regenerative or earth-healing process. An authentic eco-spirituality should be incarnational in character and based on the laws of ecology. One such ecological principle is that nature always recycles and never wastes. It follows that in areas of spiritual growth we need always conserve the resources we have at hand, seeing our collective past as a useful resource, a rich spiritual tradition. A wounded earth touches the whole social fabric and impedes us all. Recognizing this social sin may force each of us to confront our powerlessness. In isolation *I* am virtually powerless; with God's help *we* can be empowered in community and undertake the enormous task of healing our earth.

Earth healing requires a critical concentration of human and financial resources and technologies, in fact, more than ever before undertaken. These are needed for monitoring, collecting data, testing, diagnosing remedies, initiating and conducting the painstaking process of earth healing. A serious earth-healing enterprise requires instruments, materials, technical skills, and experienced personnel. Through suitable tools and application we recognize the damage done (diagnosis), we try to conserve resources for others (ecological first aid), we discover that our inexperience is costly (insight), we communicate with others to overcome barriers (consultation), and we move on to a new creation (cure). During the process we also arrive at judgments that some tools are too dangerous given their nature and the condition of the handler, for example, nuclear power generation.

People work best with suitable and adaptable tools. We shouldn't kill flies with sledge hammers or use heavy tractors where simple manual or animal-powered tillers could do the job. The technology needed for earth healing must manifest a human and earthly face, one that allows the user to grow in self-esteem and be enlivened— not deadened through routine, danger, or stress—while operating the device.

At the 1980 World Council of Churches consultation on appropriate technology we found that *appropriate* in both official languages (English and French) could mean "suitable and fitting" (the context of this discussion) and also "that which can be appropriated or made one's own." But this second definition can lead to misinterpretation, for technical imperialists may conceive of the enlightened West (or North) as having the technology that can be delivered or "appropriated" by less fortunate peoples. On the other hand, even appropriation may have a sharing or solidarity context, namely, interchanging appropriate technologies among all peoples as equal partners.

Green economics calls for locating nearby sources for bulky necessities such as food, water, fuel, and building materials in preference to distant ones. The ecological savings are immense. Thus one shifts attention to decentralized water collecting and purifying systems, dry composting toilets to reduce water use, a multitude of solar applications, and growing one's food within the neighborhood. At our Kentucky Appropriate Technology Demonstration Center we maintain a wide variety of simple and low-cost devices that reduce transportation costs and require fewer economic and environmental resources. We recognize with other appropriate technologists that shipping oil from Arabia on enormous, accident-prone tankers is risky and costly, especially when virtually free solar energy is right outside everyone's door. The challenge isn't to invent new appropriate technologies but to popularize decentralized forms such as solar energy applications.

Ideas, however, need to flow freely. They are not physically bulky and can travel with relative ease throughout the world, especially when filled with creative potential. E. F. Schumacher stated that at the level of appropriate technology "there is an almost total lack of effective international communication."[9] But such communications systems do not fit easily within the already stated criteria for appropriate technologies. Can I justify using a computer for this work and be true to what is appropriate? Computers are complex and often strain this user—and yet save time, energy, and human resources. Granted the media used to communicate ideas are complex and expensive—the communications satellite, radio, television, printed media, FAX, computer, and modem—still the potential for assisting great numbers of people makes the cost per person quite low and the resources used to make and maintain them modest. This is not true of our individualized automotive transport system. Today global communication far exceeds global commerce in uniting our world; it is powerful, dangerous, a growth industry, and something that can at a deeper stage be regarded as appropriate.

A suitable technology meets basic human demands for food, clothing, shelter, and fuel. Progressively deeper calls to perfection, faith, power, attitudes concerning the poor, and eco-justice are made on each individual journey through life.[10] Our healing as with all healing is developmental. We are not converted in an instant, but over a long period of time involving the ups and downs of human and spiritual growth. The manner in which we regard the tools of our use indicates something about this spiritual pilgrimage, and so we can reach deeper levels of technological understanding which, in turn, can act as a gauge of our own individual or communal growth.

First Stage: Technology is necessary for our survival and betterment. None of us could read this work without modern technology in some form. Our food, fuel, water, building materials, transportation of products, and communication of ideas constitute this first level of technological understanding. However, procurement of basic materials, while worthwhile, is seldom properly controlled and fraught with dangers that may lead to overuse of material things. How much is enough? When must those with more than enough start to share from the surplus accumulated? We are left dissatisfied because basic technology can lull us into acquiring and retaining control over basic commodities.

Second Stage: Realizing the effectiveness of an applied technology incites us to utilize such technologies in assisting others. We are liberated from essential local pursuits to extend a vision to a neighborhood that includes others who are in need. Tools become the means to expand outward. Unfortunately, such tools at this level are generally the ones which the person who ventures forth considers most important, and there is a tendency to impose these tools on others who may already have good ones. Their economic and physical condition may be due to other circumstances, not their choice of tools. Donors may become imperial, exert power over others, and thus apportion the tools for self-interested control and extension of influence.

Third Stage: A number of observers have pointed out the flaws of the second stage, and so we move on to a third stage of technology, a period of reexamination, critical reflection, and rediscovery. Here we focus on recognition of the limitations of the technologist. Because the tool is powerful, it can easily be misused, and so careful application is the center of attention. Because we cannot handle a device properly, we can and do damage individuals, the community, or the environment. The world is not one of black and white but shades of gray. Flawed activity gives way to a sense of powerlessness and uncertainty. It is the dark night of the technological soul.

Fourth Stage: The moment of powerlessness is critical, for we may revert to denial, excuse, or escape, and a return to previous stages. But sooner or later we must confront the flawed nature of technological change and recognize that we need one greater than ourselves to help us overcome the impasse of our journey. Technology is no savior, but it is good in its limited way. With proper assistance we can move forward, even when the assistance is from a deeper level of power that recognizes powerlessness. More subtle forms of exploitation could arise here—all the more reason for closer collaboration with a community of grace-filled people. This is the stage of global communication, when more sophisticated tools are appropriately called forth requiring collective and participative control, lest subversion of the human hearts and minds occurs to ever greater degrees.

Fifth Stage: A technology that truly serves the people and brings on the full effects of the resurrection of Christ is one that we can identify with, where the poor can be served in decency and the earth protected in the same act. Here the divine creative, redemptive, and enlivening work shows itself, a true Trinitarian action. To see its limited but essential role in saving the earth, we affirm that technology may show the unfolding Trinity at work in our world.

Technology as Enlivening the Earth

The resurrection becomes alive and moves us in a very special manner, giving us hope that victory is certain and the process has begun. However, the New Heaven and New Earth are not just the final transformation; they are already being born in this present time in which we are living and working. To heal the earth involves practical skills and expertise and invites humbling self-correctives.

To heal the earth through appropriate technologies has a down-to-earth character described in the Letter of James as well as in some of the Pauline epistles. How can we feed hungry people other than with tools to cultivate, harvest, transport to mills, grind, bake, and ship to where the needs are? All of this feeding takes technology that even the most avid anti-technologist must concede. To say "good bye and good luck," or even notify a local authority or write a congressperson, is not enough. We begin at home by better feeding our families with nutritious food that is organically grown—and abstaining where possible from the products of chemically sprayed fields. We champion the natural ways, just as we have in other methods, as church and society. We create an environmentally safe home with no harmful materials around. We reduce or eliminate the use of any foreign substances, whether in our food, home, or leisure time. We strive to find local sources of fuel and use these conservatively. We discover native building materials and fashion these into serviceable buildings.

If we move from this local level of consciousness and act accordingly, without waiting to be commanded, then we become conscientious earth healers. No individual exotic dream exceeds the hope founded in social justice that all have a decent living: wholesome food, clean water, a safe and comfortable home, proper recreation, a basic education, good health, and a good environment. How else can we make these dreams a global reality except by tackling the trillion dollar world annual defense budget, which could in great part be converted from military to social needs? These dreams focus on the poor, the key to enlivening this world and on whose side we as church have made a preferential option. Wealthier conference-goers and world travelers may feel neglected but they are asked to be humble, to start feeding the hungry, and to enter into the earth-healing process. Unfortunately, less attention is given to the dreams of the hungry and destitute than to upper-class dreams.

We often pray "give us this day our daily bread," but for those without tomorrow's bread this has special meaning. When we are fully in solidarity with the poor, they become "we" as a community, and we can again say the Our Father with meaning and sincerity. We become aware that the earth remains festering and unhealed unless the one-fifth of the globe's population of *haves* shares with the four-fifths *have-nots*. Class divisions are not perpetual. Such real divisions lead to accelerated turmoil and postponement of ecological cooperative endeavors. If and when all again become a community of believers, we are able to tackle global problems together. Failure to see, understand, and identify will only continue the divisions.

If we truly believe that all of earth's creatures are interrelated, why are we not working harder to abolish the class divisions that separate the destitute from the super-rich? Or are ecological words not truly believed? If all creatures are interrelated, what harms some will harm the rest. Social justice and eco-justice are one. If we overlook our hungry human sisters and brothers, we will most likely overlook the other creatures as well. If we are concerned about others and omit our hungry human neighbors, we have so weakened global community that peace and stability are undermined, and all will ultimately suffer from the neglect. It is not either/ or, but rather both earth and human beings.

The techniques needed to heal the earth can only do so if we give special attention to healing the users through God's good grace. This will not come about by wishful thinking; it demands hard work by a great number of people. Here we confront the dilemma, for to apply appropriate technology well should liberate people from the burden of drudgery and hard labor, yet not invite them to the irresponsible life of total leisure. Those saying no jobs are available only speak in the context of our limited economy. In fact, a multitude of meaningful but possible earth-healing jobs are out there, enough to occupy every working-age and able person. The challenge is to discover ways of making them recognized and economically viable. Much depends on converting military jobs to eco-defense ones, moving from high-paid nuclear weaponry jobs to lower but decent and satisfying ones constructing affordable homes, cleaning up the environment, rebuilding the transportation and communications infrastructure, teaching youth through formal outdoor experiences, and on and on. Redistributing wealth and resources is part of healing our planet.

Church institutions could play a vital role in introducing an appropriate technology into a global earth-healing process in a number of ways:

Discernment. People are wandering amid pulls from every side. Many seek to draw from the church's deep spiritual tradition to help overcome stress, determine their vocation, and find the best manner of praying. However, more than ever, a need exists for a social discernment to help determine what is a proper or appropriate technology. Addressing this issue requires that we return to proven ways of spiritual direction, namely prayer, abnegation, fasting, penance, and group discernment. A more recent spiritual resource is the small and prayerful base communities, composed of like-minded people who are committed to working out problems together. They neither waste time distributing false accolades nor wilt at the prospect of environmental doom.

Linkages. The Catholic church network makes it an ideal global linkage for appropriate technology. Certainly other global groups, such as multinational corporations, do not meet the need for appropriate technologies as such technologies are generally not highly profitable. Furthermore, the church is committed to networking and building community. Solar cookers need to replace expensive kerosene units, but adapting to local customs may require sensitivity on the part of church workers. Adaptation and inculturation are becoming sufficiently understood, and a sensitivity is growing that will identify and restrain some forms of technological imperialism found in the past. The church needs to be at the cutting edge and initiate programs that will be incorporated into governmental agencies' operations.

Liturgical Participation. The church's liturgy trains participants to value the gifts of one another, to see that Martha and Mary need each other for a viable community. The mix of people is further enforced by the rhythm of the celebration, where liturgical word and deed touch devout hearts. Meaningful public liturgies need to invite people of all different ranks, cultural backgrounds, and occupations. Connection with the poor is a self-corrective process, for it rids the institution of that which is superficial and opens it to a more down-to-earth approach to service.

Ecological Facilities. The church's physical facilities need to show the caring for creation that is spoken of in scripture. The structures should be energy-conserving both in heating and cooling, non-wasteful in building materials, safe, affordable, and, very important, beautiful. Grounds should be tastefully designed with a proper ornamental and utilitarian balance, so that vegetation acts as a sound and privacy barrier and furnishes fruit and nuts, shade, and protection from winter wind. All such facilities should have environmental assessments before basic construction or additions are begun to permit good ecological design.[11]

Challenge Attitudes. The place of the appropriate technologists and engineers in helping to solve the environmental crisis must be acknowledged by giving them a proper role in catechesis involving ecology. Appropriate technology has not been fully endorsed either by the environmental community or by governmental decision-makers, who often relegate it to the work of tinkerers, eccentrics, and inventors. Often professional environmentalists link with regulators to define the "real" issues to be smokestack filters, sewer systems, and the billion dollar appropriations stuff, while neglecting appropriate technologies. Changes will not occur until programs of catechesis on all levels are initiated, and here the institutional church can take a leadership role.

Advocacy. Church groups are often ideally placed to advocate for appropriate technology methods that incidentally could include home gardening methods, plentiful clothing materials, renewable energy fuels, affordable health care facilities and educational opportunities, lower-priced nutritious and natural foodstuffs, super low-cost housing, and even natural forms of birth control. Financially harder times require that governmental money be targeted for the poor and that the religious community inspire those affected to translate their legitimate needs into such political activities as lobbying, supporting the proper candidates, and following the course of legislation.

Demonstration. Some of the best examples of Catholics working quietly in the environment are the sisters' and laywomen's communities in many parts of the nation; they are the bright lights in the American environmental movement. The church's contribution has been in its tireless work in farm, retirement home, and motherhouse. It is found in herbal and organic gardens, seasonal extenders, solar greenhouses, windmills, composting bins, managed forestlands, wholesome meals with reduced use of meat, crafts and skills of native peoples, vegetative barriers for shade and visual screening, wildscape, and edible landscaping.

Female and Male Spirituality. Healing has often been seen as a female calling. Broadening the church's ministry often takes on efforts to expand ministry to women. However, males do not take easily to better housekeeping, and that includes ecological concerns, the ultimate housekeeping. The church has long defended the maleness of certain ministries; however, now it must gently and firmly reconsider certain roles, and for this leadership we need the total church, females and males working together. Women healers will play a great part in the total earth-healing process.

Conclusion

The church working in collaboration with all people of goodwill must hasten the day that appropriate technology is fully incorporated in healing the earth.[12] As a priest I find a more ministerial role in developing a better dry-composting toilet than attempting to craft the final word on eco-spirituality. In fact, at this time when appropriate technology is denigrated, a better form of eco-spirituality may be the redemption and advocacy of an impoverished technology that arises from the poor and is capable of helping the poor—people and earth. Isn't catalyzing this process precisely the role of our servant and evangelizing church?

Holy Mother Church calls us to be good homemakers, for we are asked to prepare our abode for the Second Coming of the risen and glorious One. Servants tidy the house for the long-awaited guest, whose advent is certain and whose coming must find us ready and waiting. To clean up involves using the proper instruments and methods, and that is part of the readiness. Those unconcerned about either spreading or awaiting the good news may find any discussion of types of technology distracting. However, we are a people incarnating the spirit of God, and matter enters into our spiritual considerations. With expectancy we continue to ask whether a particular technology is necessary, humanly fulfilling, and revealing of the Trinity in our midst. Such an ongoing critique will undoubtedly advance the cause of earth healing and prepare for the day of the Second Coming (2 Pt 3:11-13).

Notes

1. Ken Darrow and Mike Saxenian, *Appropriate Technology Sourcebook* (Stanford, Cal.: Volunteers in Asia Publications, 1986), p. 7.

2. Nicholas Zernov, *Eastern Christendom* (New York: Putnam, 1961).

3. Lionel Casson, "Godliness and Work," *Science* 81, pp. 36-42.

4. Ibid.

5. A more lengthy discussion is given in Albert J. Fritsch, S.J., *Renew the Face of the Earth* (Chicago: Loyola Press, 1987), chap. 7.

6. Lynn White, Jr., *Science* (1967).

7. "Renewing the Earth: An Invitation to Reflection and Action on Environment in Light of Catholic Social Teaching" (U.S. Catholic Conference, November 14, 1991), p. 7.

8. Edward Echlin, "Dare Ecology Use the Word 'Sin'?" *The Month* (May 1993), p. 206.

9. E. F. Schumacher, *Appropriate Technology Journal* (1976) quoted in Darrow and Saxenian, p. 6.

10. See Bob Sears and Albert Fritsch, *Earth Healing: A Resurrection Centered Spirituality*, in press. Copies of the manuscript can be obtained by writing to ASPI Publications, P.O. Box 298, Livingston, KY 40445.

11. *Eco-Church: An Action Manual* (San Jose: Resource Publications, 1992).

12. Pope John Paul II, "The Ecological Crisis: A Common Responsibility" (1990), no. 15. The pope says that the ecological crisis has assumed such proportions as to be the responsibility of everyone.

9

Eating the Body of the Lord

Eucharist and Community-Supported Farming

MARC BOUCHER-COLBERT

Since I first read them, these words of the poet William Carlos Williams have worked their way into my vision of Catholicism, specifically of the eucharist:

> There is nothing to eat,
> seek it where you will,
> but the body of the Lord.
> The blessed plants
> and the sea, yield it
> to the imagination
> intact.[1]

This poetic phrase resonates with the weekly proclamation of the eucharistic minister at the distribution of communion: "The body of Christ." Yet it also amplifies our sense of Christ's body by drawing attention to the whole created order. The quotation was for me the first in a series of clues of how to reinvigorate the eucharist, not merely the ritual but the whole celebrating assembly, and make it meaningfully profess Christ's resurrection in the face of ecological despair. The path has led me to community-supported farming, a radical new approach to agriculture and a fitting earth-centered complement to eucharist. In my thinking and experience I have come to see community-supported agriculture as a sacramental encounter with the body of Christ in its human and cosmic dimensions. This chapter elaborates on the reciprocity I see between eucharist and agriculture.

As a primary symbol of the church and as the primary focal point for the life of believing Catholics, the eucharist must necessarily come into contact with the problems and troubles of our time and culture; not only that, it promises new hope in the midst of suffering and perplexity, for it is the means par excellence by which the community of faith encounters Christ present in its midst, a presence heralding and embodying God's transformative love.

If there exists a "deep seated persuasion that it [eucharist] is the faith-community's central confession of faith and the key moment of the presence of Christ and of the Spirit in the church," there is also the sense that its power to "offer an overarching vision of the world" has broken down.[2] So many of people's anxieties, outrages, and grief occasioned by the modern world appear unable to be incorporated into the matrix of the eucharist; thus excluded, they remain untransformed. Chief among these is the degradation of our planet affected by the modern industrial-technological lifestyle.

Gathering weekly to partake in the Paschal Mystery, celebrating the victory of life over death, and then going forth routinely to destroy the water, air, soil, and creatures of the earth is a contradiction that is at best hypocritical and at worst deeply pathological. Currently, the church seems barely conscious of this schism between its profession and its practice. Can the eucharist be the means Catholics use for transforming the earth in God's image, or does it speak to such a narrow reality within and among us that in the face of crises, it will be reduced to pious trivia? Surely the salvation Catholics celebrate in Christ must shed some light on the impending ecological disaster. Otherwise we are relegating Christ to that "millennium" at the end of history, a mythos to which Thomas Berry attributed all the worst aspects of the present state of affairs.[3]

Searching the eucharist for a source of grace to overcome planetary degradation is by no means a coincidence. We can look deeply into this sacrament because of its nature: eating bread and drinking wine, gifts of the earth which bear the sacramental presence of Jesus Christ. Historically, too, the eucharistic celebration is rooted in the Jewish Passover and matzo feasts, which hearken back to earlier agricultural festivals of spring lambing and the grain harvest. That these agricultural gifts and motifs have been understood to reveal divinity so profoundly attunes us to the sacramental power of the earth. Yet with our electric lights, processed foods, and condominium developments, we are all but cut off from the divinity shining from our natural surroundings. Perhaps our divergence from the powers and rhythms of nature finds its appropriate liturgical symbolism in the flat, bland, communion wafer that is supposed to be bread. Our liturgies, like our lives, are radically disconnected from the divine revelation in the universe. We have reached a juncture in our sacramental life where, in order to save the created order, we will at least have to allow it a privileged place among our symbols for divinity. A sacramental *perestroika* is in order, and this entails a return to sacramental basics.

In his primer on sacraments, Tad Guzie describes a rhythm by which sacramental experience develops. He calls it the "rhythm that makes life human": lived experience, story, festivity.[4] Sacraments begin from our everyday lives when we are attentive to life's meaning and when we are connected to a community with which to share that meaning. All human communities have experienced sacraments, in the general sense of that term, insofar as they were attuned to the mysterious, transcendent, or precious aspects of their experience.[5] When we reflect on our experiences and find deeper meaning in them, and eventually share that meaning with others through story and festivity, then we too are living the sacramental life. In this sense we can see the sacraments as truly human actions not something imposed by God from outside the human situation or something arbitrarily commanded by Jesus. And yet sacraments are divine actions, too. Within the human situation God touches us and draws us more deeply into meaning, mystery, and love. The divine Mystery actively allures us. Sacraments tell the story of our meeting with God and celebrate the encounter through ritual and symbol.

In the Catholic church, however, we are aware of a more restrictive definition of the word *sacrament*. In light of history, we know that from the twelfth century onward, the church has committed itself to seven master symbols or rituals, which are privileged expressions of God's activity in the world, as seen through Jesus Christ. These seven sacraments share similarities with the sacred rites of other human communities; they are rituals of initiation, sacrifice, atonement, commitment, and healing. The gospels bring to these human actions a fuller meaning, and they become vehicles for meeting a God with whom we can personally relate. Yet in the historical development of the Catholic sacraments, layers of theorizing about the meaning and effectiveness of the rituals have removed them from the level of immediate experience for most believers. In addition, "the sacraments have become so 'churchy,' so separated from the lives we live at home and at work and at play, that we no longer spontaneously relate them to the other symbols that surround and affect us."[6] Many Catholics no longer see their religious practice as a vital part of *real* life; this results in the loss of the transformative power of the sacraments, with which we might change the culture and restore the planet.

The power of sacraments comes in part from our experiential contact with the symbolic elements. Symbols, while retaining their own meaning as objects or actions, also lead us to a deeper level of meaning or experience; symbols are our doors to the sacred. Ironically, even though our culture is drenched in images, we moderns lack a highly developed symbolic imagination. The general impoverishment of symbolic thought has taken its toll on church life too. Consider the following shift in the symbolic action around the sacrament of confirmation: "Confirmation began as a rubbing from head to toe with perfumed oil, and what an experience it must have been for the assembly when the newly baptized came into their midst, truly exuding the sweet fragrance of Christ. As for today, it is hard to see what human or divine significance can be read into a flick of odorless oil on the forehead."[7]

When we reflect on eucharist, at its very heart we find an intimate meal. Yet how often does Sunday liturgy remind us of a meal? While we have been taught to view this event from the lenses of covenant, sacrifice, and Paschal Mystery, among others, the root experience is eating together. "The element of 'communion' means that the eucharist is a meal, a community meal, in which all the participants are brought together to have a common share in common goods, these common goods being first of all the bread and the wine of a real human meal, whatever their deeper significance."[8] If food and eating are the key symbols of the eucharist, then we need to probe their significance more deeply. If we involve ourselves with the "gifts of the earth and work of human hands" that the bread and wine symbolize, then we will have begun to reinvigorate the eucharist.

A friend has suggested that a complementary chapter could be written taking as its starting point the question: How do we eat an intimate meal together? He has suggested Jean Vanier's material on community as a foundation for the answer. The point is well taken; it is bread broken and wine poured out that symbolize Jesus. For our part, we are seeking an answer to the question of how the eucharist might transform the current degradation of our planet, and so our frame of reference will be more biocentric, focused on the food itself.

Wendell Berry, in a short essay entitled "The Pleasures of Eating," acknowledges that eating is part of a larger agricultural drama.[9] It is within that drama that I want to continue to probe for revitalizing symbols; for I believe that our response to the divine mystery of a meal has atrophied due to our loss of connection with the fruiting earth. Because of this loss of intimacy, the power of a sacrament of eating and drinking has dimmed. Also, given the eucharist's agricultural origins, it seems appropriate to proceed in this direction. Both the symbols and history of this sacrament, then, lead us to reflect on agriculture.

In ancient societies agriculture was always regarded as a sacred act and accompanied in its varied stages by appropriate rituals. In more recent times, following the desacralization of the natural world, farmers, if not seen as actors in a sacred drama, were thought to be guardians or custodians of important resources and knowledge. Their contact with the earth was held to be a most intimate and careful one, worth pursuing from one generation to the next. Farmers "grounded" the society and reminded people of the greater rhythms on which their lives depended. Even city dwellers in the earlier part of this century were not far removed from local farmers, and those who were not directly involved in their own food production probably were aware of the people who were. Food and eating were not detached from their sources by excessive transportation, processing, or packaging.

Until fairly recently, most consumers were more modest, more aware of their role as producers responsible to meet many of their own household needs.[10] Consider the fact that we have available to us only 3 percent of the kinds of food plants that existed in our grandparents' day.[11] To me this indicates a tremendous loss in the care and propagation of species as well as loss in the value of producing our own food. Gardening has become more a hobby, chosen for its relaxing benefits, than an integral part of the lifestyle of a home economy, and industry has stepped in to replace household effort. Its standards have been "high yield, ease of shipping, cosmetic appearance, and uniform ripening," not nutrition and taste, and our food and eating have declined along with our health.[12]

Agriculture has leaped from small scale to a global-industrial system that has "not only destroyed thousands of rural communities but, through its reliance on pesticides, animal drugs, and chemical food additives, is gradually destroying our health. The system knows no national boundaries, nor does it respect cultural traditions, community bonds, human health, or the sacredness of life."[13] Agriculture has yielded up a sacred standard—the measure of life—to profane and diabolical ones, in their lack of sensitivity and limits—the measures of the machine and the market.[14] Food production has become the province of business people as much as of farmers—to its detriment. Farming as business involves the manipulation of plants and animals to maximize production and profit; science, the henchman of agribusiness, furnishes the necessary technology to continue the assault on nature's limits and to increase productivity. Yet the farmer's true art, now almost lost, attempts to understand natural forces so as to work with them and within their boundaries.

If farming was once sacramental, our materialistic and consumerist vision effectively has screened the divine from view. The food we eat is, by and large, no longer the product of a caring, perhaps even religious encounter with the forces of nature. We are eating the fruits of destruction and exploitation. No wonder eucharist has so little power to change us; our one sacred meal a week comes up against twenty-one or more unholy ones, not to mention snacks! As farmers have fled the land due to economic hardship, as food has become poisoned with chemicals and disease, and as the very soil has been lost through erosion, we can only lament what has become of our most basic and life-sustaining relationship with the earth.

The crisis, while tragic and still unfinished, has not been without any hope for agriculture. Beginning in Europe and now spreading rapidly across the United States small community-supported farms have emerged which promise to restore people's vital connection to the earth through agriculture. A community-supported farm consists of a group of people who band together to maintain a small farm, which in turn supports them with fresh food. All of the community-supported farms with which I am acquainted practice organic growing techniques, replacing the use of chemical pesticides, herbicides, and fertilizers with alternatives that often emphasize low-tech, labor-intensive practices and always insist upon an overall contribution to the soil's health.

The dynamics of a community-supported farm differ dramatically from a market-oriented farm. A few of the members most skilled in farming take the responsibility for the full-time work, while the others meet the farm's expenses with a financial contribution. In general, members contribute their share of the cost in advance of the growing season, risking with the farmer any hardships the weather, pests, or disease might bring. Payment in advance also reinforces the members' support of the entire farm and prevents the "cash-for-vegetable" transactions that feed the consumer mentality. Their contribution entitles members to an equal share of all the farm's bounty. Members eat produce in season and may be called upon to help plant, harvest, or tend the crops, thus entering into the life cycle of the growing things on which they depend. They interact with farm animals and learn their behaviors and temperaments. They celebrate, with the farmers, the various festivals of the growing season.

The community-supported farm transcends a mere human economic arrangement; it involves all the creatures and forces of nature, the larger community of which humanity is a part. Trauger Groh, a pioneer in the community-supported farming movement, relies on the leading concept of the farm as an organism to guide his vision of community-supported farming. Not all community-supported farms practice this organic approach to farming. Many do not even have animals, which make the farm organism possible, due to size restrictions or focus. The farm-as-organism concept, however, provides an ideal state toward which farms can move and illustrates the self-renewing powers of the earth on a microcosmic level.

This concept, generated by Rudolf Steiner, the Austrian philosopher and natural scientist, calls for the farm to be characterized by distinct boundaries, which mark its unique identity, and for an inner life and circulation that is different from its surroundings; in other words, the farm should become an organism. In this view the cooperation of domestic animals is essential. "Ideally, they feed from the vegetation inside the farm. . . . The animals respond with a manure that is formed exclusively by the flora of the farm organism. The manure is collected and, when properly treated, comes back to the plants of this place, stimulating them. In this process of correspondence between farm animal and farm vegetation, the farm develops and becomes more individualized. An organism is born and develops in time."[15]

Attention is also given to the development of wetlands and hedgerows in the farm, permitting a certain wildness to flourish within the domesticated boundaries.[16] Wendell Berry's thought reinforces the importance of these margins. He praises Andean agriculture for including margins in its agriculture: "The hedgerows are marginal areas, little thoroughfares of wilderness closely crisscrossing the farmland, and in them agriculture is constantly renewing itself."[17] In its largest scope, the community-supported farm, a body of domestication, is in constant interaction with itself and with the wilderness on which it depends. It is a human and animal and plant and mineral community, supported by the life forces surging through it all and greater than the sum of its parts.

My plea for a deeper connection between the eucharist and farming is no mere academic exercise. I am working out this chapter in the context of my life as both a practicing Catholic and a community-supported farmer. My experiences from the farm have convinced me that people do experience farming as a sacramental reality, especially people who have had little contact with the earth and with living plants. Our farm, called Urban Bounty Farm, is located within the city limits of Portland. We feed twenty-seven families or households on three-fourths of an acre. We ask our members to come to the farm each week to pick up their produce, because we feel the contact the members have with the land, with the active farmers, and with one another has immeasurable benefit in reestablishing their link with the natural world. The looks of wonder and delight on members' faces—adult and child alike—as they discover an eggplant or as they sample a succulent and sweet cauliflower, so different from the supermarket offering, are typical of the awe that accompanies sacramental moments. To some degree, as with sacramental rituals, when members come to the farm they enter another space and time.

To arrive at our farm a person must drop down from a busy road to a secluded creek bottom. On driving down the gravel access road, one passes through a thick, shaded stand of fir trees, which quiets and protects the canyon and completely obscures the road above. Emerging from the trees and thick brush at the creek level, a sunny field full of the resplendent greens of a variety of vegetables meets the eye. A real transition has been made, similar, though not as intense, to the rite of passage that necessarily precedes many sacramental encounters. The space is the ordered and pleasing array of a large garden, surrounded by the wildness of the forest on the canyon wall to the south and east, a cedar hedge to the west, and the creek to the north. It is the home of birds, small animals, and the vegetables—a space comfortable for humans but not designed only for them. And in the peace of the place, our members and visitors noticeably relax.

Time also changes for those on the farm. Contrasted to the relentless pace of living by the clock, the pace of the farm is more dictated by the heat and light of the sun and the cycles of the vegetables. My farming partner and I work hardest in the cool of the morning and evening. We take a nap at midday to avoid the heat. When members come during the late afternoon and early evening of their designated pick-up day, the setting sun casts a radiant glow over the entire place and the air has cooled. People tend to linger and act unrushed. They become aware of how the vegetables are faring, how far along they are to flowering or fruiting or harvest.

People can sense the seasonal rhythms more deeply by becoming aware that certain vegetables come and go with the seasons; beautiful, bluish heads of broccoli and sweet snap peas mean the cool, wet spring is still here; tomatoes, basil, eggplants, and peppers are the heralds of hot summer, along with the explosive, firework-like summer squash. The return of leafy greens, the bulging of beets below the earth, and the coloring of pumpkins announce the fall. And the proud, erect Brussels sprouts and the stout cabbages signal the earth's providence, even in winter's death.

The following description of a sacrament fits my own experience and the experiences of many I know with regard to our community-supported farm, and for that matter, many other farms. Sacraments "occasion a transformation of consciousness, an alteration of a person's experience of space and time, in which things become imbued with new meaning and value."[18] When people are in contact with their food in the larger context of a farm, they can experience the sacramental realization that "we are living from mystery, from creatures we did not make and powers we cannot comprehend."[19] In the supermarket, under sallow lighting, cut off from its context, food loses its mysterious radiance.

From even this quick survey of community-supported farming, comparisons with eucharist come easily. The eucharist, in the broader sense of liturgy, is a repetitive sacrament whose context is, in its origins, the planet's own rhythms. The major feasts of Christianity—Christmas (celebrating the mystery of the incarnation of God) and Easter (celebrating the resurrection of Jesus Christ from the dead)—both stem from the earth's and sun's rhythms around solstice and equinox. Because of liturgy's internal rhythm, which harmonizes with the earth's own, eucharist is already a cosmological sacrament. Yet anthropocentric tendencies have limited the fullest development of this cosmic attunement. The shift in focus from the blessed sacrament (the bread and wine as body and blood of Christ) as an object of worship, to the entire liturgy, including the community gathered, as an act of worship, has been welcome.[20] Yet it still fails to embrace in a meaningful way the whole natural community and the deep mysteries of the earth itself.

In a similar way, the grand cathedrals of Medieval Europe represent a stage of heightened cosmic awareness in the Christian religion. They are, almost literally, universes of images, meaning, and mystery, integrating Christian and natural symbolism, into which the believer is plunged in the midst of his or her sacramental celebration. The loss of such rich cosmology in the modern world has impoverished the eucharist and robbed it of a living connection to the cosmos, with which it might renew the world.

Yet in both the liturgical year and the cosmic sensitivity of the past, we see the seeds of hope that eucharist might speak to the current human assault on the planet. People to whom I've mentioned the connection between eucharist and farming have often wondered aloud, with a puzzled look, how such things could share any commonality. Theologically, the church has failed to develop a model for understanding the relationship of God's personal revelation in Jesus with the revelation of God in all the created order, but its failure to provide such a model does not mean these resources do not exist within our tradition. Matthew Fox has recovered the tradition of the Cosmic Christ from its near banishment at the hands of Enlightenment forces of rationalism, tracing its presence in the wisdom tradition of ancient Israel, in its prophetic heritage, and in its apocalyptic literature. He continues to follow its development in the Christian scriptures as it is applied to Jesus, through the Greek patriarchs of the early church, and into the Medieval West, where such cosmic awareness flourished from the twelfth to the fifteenth centuries. This Cosmic Christ tradition, in my opinion, offers Catholics and all other Christians the key to rejoining our worship with the liturgy of nature.

In ancient Israel's awareness, pre-existent Wisdom, which is personified, was thought to pervade all things and to have brought all things into being. The author of the book of Wisdom, speaking in the person of King Solomon, says this:

> For he gave me sound knowledge of existing
>   things,
> that I might know the organization of the
>   universe and the force of its elements,
> The beginning and the end and the midpoint
>   of times,
> the changes in the sun's course and the
>   variations of the seasons.
> Cycles of years, positions of the stars,
> natures of animals, tempers of beasts,
> Powers of winds and thoughts of men,
> uses of plants and virtues of roots—
> Such things as are hidden I learned and such
>   as are plain;
> for Wisdom, the artificer of all, taught me.
>   (Wis 7:17-22, *NAB*)

From this perspective, there is no sharp distinction between the knowledge of the universe and the knowledge of God. In seeking Wisdom, the writer sought the very power by which God established and governed the universe.

In their post-resurrection experience of Jesus, early Christians continued this strain of thinking as they discerned cosmic implications in Jesus's life and mission. The prologue to the gospel of John "celebrates Christ as the Logos or 'Word' of God as well as the wisdom of God present with God at the creation of the world"[21]:

> In the beginning was the Word:
> the Word was with God
> and the Word was God.
> He was with God in the beginning.
> Through him all things came into being,
> not one thing came into being except through
>   him. (Jn 1:1-3, *NJB*)

And in the writings of Paul, ancient hymns of Jesus's cosmic significance abound, for example:

> He is the image of the unseen God,
> the first-born of all creation,

for in him were created all things
in heaven and on earth:
everything visible and invisible,
thrones, ruling forces, sovereignties,
    powers—
all things were created through him and for
    him.
He exists before all things
and in him all things hold together.
       (Col 1:15-17, *NJB*)

These texts emphasize the transcendent nature of Jesus Christ and allow us to see in the historical figure Jesus of Nazareth, in his ministry and death and resurrection, the key to understanding the mystery of the universe. Conversely, in light of this cosmological framework, we can also roam beyond the confines of traditionally sanctioned religion and expect to see Christ. Our whole lives can be experienced as sacramental, given that we reflect on them and discern the Wisdom, the Logos, the Cosmic Christ at the center of all we experience.

After tracing the path of cosmic Christology through Western culture, Fox summarizes the "signatures" of this Cosmic Christ, the ways in which we recognize this divine manifestation. A sense of pattern and coherence emerges as the Cosmic Christ's trademark. The Cosmic Christ gives a cosmic purpose to our lives and imbues the entire universe with a new meaning; the Cosmic Christ joins mysticism, art, and science in the search for God; the cosmic dimension of Christ is holistic, connecting microcosm and macrocosm.[22] All these attributes of the Cosmic Christ can be learned in a very direct and experiential way from farming, making the farm, especially a community-supported farm, an ideal training ground for a cosmic religious awakening, for a sense of eucharist that includes all of creation.

As I have mentioned before, the farm, when properly developed, can be thought of as an organism, a holistic reality. It is a body of Christ in the cosmic sense just as surely as the community of believers is the continuing body of the resurrected Jesus Christ. In its wholeness, the farm embodies the unity, beauty, and coherence of the whole universe. Through the complex interplay between the domesticated animals and plants and their wild companions of the hedgerows, forests, and other margins, one observes in the interactions of the farm as microcosm the more complex and mysteriously ordered macrocosm. If contemplated at all, these processes are likely to arouse awe and wonder; if studied deeply, they may become the foundation for an advanced mysticism. Says artist M.

C. Richards: "Agriculture is a schooling for spiritual perception. For it is there that we learn to practice meditative attention: the detailed observation of the living process."[23] Just as in the eucharist we meditate on Jesus, the human face of God, so on the farm we look long and lovingly on the natural face of God.

Community-supported farming unites people in a common body, a further extension of the body of the farm. Members who pursue the art of farming become empowered to serve people at the level of their most basic needs for clean, healthy food and for the skills to domesticate and live from the earth while increasing its fertility. When the disciples, confronted with multitudes who had come to hear Jesus, panicked because they saw the people's need for food, Jesus said to them, "You give them something to eat" (Mk 6:37, *NAB*). Modern disciples who wish to follow Jesus's passion for justice and his compassionate love would do well to heed his words and take to farming. To the traditional "works of mercy," corporal and spiritual, one might add the following: grow food in a healthy way. This practice, at once restorative of body and spirit, is a form of universal service not only to our human brothers and sisters but to every creature and being in the web of existence.

For Christians, Jesus historically grounds and personalizes our sense of the divine mystery. God has spoken with a human voice, has touched us with human hands. Yet mere devotion to this historical Jesus has not engendered the psychic energy among believers radically to challenge the destructive forces being unleashed upon the planet. When we experience the divinity in Jesus that preexisted his earthly ministry and that reigns now, sustaining all creation, then we have tapped into the mystery of the Cosmic Christ, and we can celebrate God's presence among us everywhere and in everything. While we can proclaim the cosmic dimension of Jesus Christ in our churches, it is best discovered in the world of day-to-day life, most especially in the natural world. And agriculture is our most intimate connection with nature, our most constantly nurtured link. Since agriculture has become perverted, made to serve profit motives which are ruining the earth, we as Christians, in the name of the Cosmic Christ, must take it back. When people take responsibility for farming through community-supported agriculture, they are truly embracing the body of Christ, which extends through all of creation. When churches make agriculture one of their ministries, then the eucharist will be a true communion with all of creation.

As Wendell Berry observes, eating is more than a merely sensory or appetitive stimulation. It is the last act of an agricultural drama, and for him, the quality of eating increases with one's participation in the drama.[24] I draw the same conclusions for eucharist, and go beyond Berry to say that eating is a penultimate act; agriculture finds its fulfillment when we see God's generosity in our food and share that goodness with others. Sacramental eating fulfills the agricultural drama, and that sacramentality of food is crucial to a eucharist that can heal the earth. Just as we could not baptize with sewage and expect the appropriate sacramental effectiveness, so too we cannot poison our land and our food or remain aloof from agriculture and expect to find a profound ecological meaning in the eucharist.

Our symbols need to come closer to our lives. We must learn to care about the bread and wine that symbolize all the earth's gifts, our only food. Community-supported farming draws us close to the body of the Lord, the mysteries of the Cosmic Christ in all its phases, so that our liturgies concentrate and make potent the Christic food we eat daily, physically and imaginatively. Without this connection, our sacrament becomes like our eating, detached from its source, and we become malnourished as a result, too weak to fight for the defense of the earth.

Notes

1. Quoted in Wendell Berry, "The Pleasures of Eating," in *Our Sustainable Table*, ed. Robert Clark (San Francisco: North Point Press, 1990), p. 131.

2. David N. Power, *The Eucharistic Mystery: Revitalizing the Tradition* (New York: Crossroad, 1992), p. 8.

3. Thomas Berry, *The Dream of the Earth* (San Francisco: Sierra Club Books, 1988), p. 28.

4. Tad Guzie, *The Book of Sacramental Basics* (Ramsey, N.J.: Paulist Press, 1981), p. 18.

5. Joseph Martos, *Doors to the Sacred* (Garden City, N.Y.: Image Books, 1982), p. 16.

6. Guzie, p. 46.

7. Guzie, p. 132.

8. Louis Boyer, quoted in Horton Davies, *Bread of Life and Cup of Joy: Newer Ecumenical Perspectives on the Eucharist* (Grand Rapids, Mich.: Wm. B. Eerdmans, 1992), p. 119.

9. W. Berry, p. 125.

10. Wendell Berry, *The Unsettling of America: Culture and Agriculture* (New York: Avon Books, 1977), p. 24.

11. Kenny Ausubel, "Avant-Gardening: Planting Seeds of Change," in *Seeds of Change Organic Seed Catalog* (1992), p. 2.

12. Ausubel, p. 2.

13. Kathryn Collmer, "From Hand to Mouth," *Sojourners* (June 1993), p. 18.

14. W. Berry, *The Unsettling of America*, p. 89.

15. Trauger Groh with Steven S. H. McFadden, *Farms of Tomorrow* (Kimberton, Penn.: Biodynamic Farming and Gardening Association, Inc., 1990), p. 21.

16. Groh, p. 36.

17. W. Berry, *The Unsettling of America*, p. 178.

18. Martos, p. 16.

19. W. Berry, "The Pleasures of Eating," p. 131.

20. Martos, p. 301.

21. Matthew Fox, *The Coming of the Cosmic Christ* (San Francisco: Harper and Row, 1988), p. 94.

22. Fox, pp. 131-44.

23. M. C. Richards, "The Renewal of Art through Agriculture," *Creation Spirituality* (March/April 1993), p. 13.

24. W. Berry, "The Pleasures of Eating," p. 125.

10

Christianity and the Creation

A Franciscan Speaks to Franciscans

RICHARD ROHR, O.F.M.

Creation Spirituality: A Franciscan Insight

It's wonderful to be able to talk to my own community about an area that means so much to all of us. I think we'd be lying if we said that this is a positive period. We're certainly aware of great pain in the local church; we're aware of great pain in religious life; we're aware of great polarizations, great pain, in the institutional church. So, it's probably of utmost importance that we ask ourselves, "Why do we believe and what do we believe?" What is it that's good about the Franciscan interpretation, the Franciscan tradition, of the gospel?

Only insofar as Franciscanism is a rediscovery of the gospel is it of any enduring value to the world. Only insofar as Franciscanism is a rediscovery of Jesus, for each age, does it offer something enduring to each age.

First of all, Franciscans are people of the gospel, people of Jesus. That's all that Francis wanted to be. Francis was not a Franciscan. He was a follower of Jesus, and that's primarily what we all are about too.

We're living in an exciting time in terms of what's happening to scripture. We're learning to be honest about the historical, political, and economic setting in which Jesus first spoke his words and in which the books of the Bible were written. And, although a little slower, the same thing is happening in Franciscanism. We sometimes speak of "birdbath Franciscanism," meaning the overly sweet, unreal, disconnected from history and politics version of Francis that many of us have—the sweet Francis in the garden, where there

129

are flowers and birds, but the rest of the world is conveniently cordoned off. That isn't fair to Francis, and it certainly isn't fair to Jesus. Francis lived in a very real socio-economic and cultural setting, and part of the problem is that we tried to take him out of it. Soon his statues were trimmed in gold leaf, and we began to idealize him instead of imitate him.

The creation spirituality of Francis seldom has been practiced by his followers. None of the subsequent Franciscans would be known for creationism. It was almost lost immediately, because, as we all know, Francis, this amazing gift of grace, this amazing personality, held together in one lifetime and in one life's experience an amazingly broad reality. Franciscans became known for many other good things, many other gospel things: We became teachers, we moved into the universities, we became great apostles, great missionaries, known for various kinds of ascetical lifestyles. We were often canonized. But there have been few Franciscans who have been lovers of the world in precisely the way Francis was.

In the great sacred mythologies and stories of almost all the world's religions, the heroes and heroines of faith were not gods; rather, they were human beings walking a journey. Honest commentators would not be afraid to admit that some of Francis's initial responses to his own father, for example, were less than mature, less than a liberated, fully gospel response. In the past we were unable to accept him as a man who had to grow up, face his own demons, face his own dark side, repent of his own sin. But it was in that confrontation that he grew into the holy man we now call St. Francis.

Actually we're doing the church, the gospel, and certainly Jesus, a much greater service when we don't try to make the saints into gods—or try to make Franciscanism into some kind of unreal mysticism unconnected to the world as it is. Invariably, in every age and in every century and in every culture, we face the same recurring human demons and human problems. The one we'd like to address here relates to a theme that is now being called *creation spirituality*. Creation spirituality has its origins in the Hebrew scriptures, as evidenced in psalms 104 and 148, which could be another form of the Canticle of the Creatures. It is a spirituality that is rooted, first of all, in nature and in experience and in the world as it is. This rich Hebrew spirituality formed the mind and heart of Jesus of Nazareth.

Maybe we don't feel the impact of that until we realize that many people think religion has to do with ideas and concepts and formulas from books. In fact, an awful lot of clergy do. That's how we were trained for years. We went away, not into a world of nature and silence and primal relationships but into a world of books. All religion was supposedly found in books. Well, that's not biblical spirituality, and that's not where religion begins. It begins in observing "what is." Paul says, "Ever since the creation of the world, the invisible existence of God and his everlasting power have been clearly seen by the mind's understanding of created things" (Rom 1:20). We know God through the things that God has made. The first foundation of any true religious seeing is, quite simply, *learning how to see and love what is*. The contemplative insight is, first of all, learning how to see. Contemplation is meeting reality in its most simple and direct form unjudged, unexplained, and uncontrolled!

A great disappointment to many people is that institutional religion invariably tries to do an "end-run" around the body and around the soul and around creation to get to Spirit. It doesn't work. It results in very shallow religion—people trapped in legalism, in ritualism. It creates people with all kinds of conclusions, but no journey; all kinds of certitudes, but no wisdom; all kinds of supposed clarity, but no real understanding and hardly even compassion. The true religious journey is not a journey "around" but a journey "through." I mistrust people who say they love Jesus but don't know how to love a rock. If we don't know how to love what's right in front of us, then we don't know how to see what is. So we must start with a stone! We move from the stone to the plant world and learn how to appreciate growing things and see God in them, see the *Vestigia Dei*, the fingerprints or footprints of God. Then, if we pass the second grade, maybe we can go on to the third grade and learn how to love animals and learn to recognize that each one is another image of the divine.

Recently I went to San Rufino Cathedral where Francis was baptized. This apparently tenth-eleventh century cathedral has one image of Jesus right above the main portal, but the entire rest of the facade of the cathedral is animals. Animals! I wonder how often Francis looked at that facade—and what connections he made.

Perhaps once we can see God in animals, we might learn to see God in our neighbor. And then we might learn to love the world. And then, when all of that loving has taken place, when all of that seeing has happened, when such people come up to me and tell me they love Jesus, I'll believe it! They're capable of loving Jesus. The soul is prepared. The soul is freed, and it's learned how to see and how to receive and how to move in and how to move out from itself. Such individuals might well understand how to love God.

Now that, quite simply, is creation spirituality. It's not from above, but from below. It begins with what is. The primary revelation of God is creation. Maybe this creation has been around for twenty million years, maybe a lot more than that. God has been speaking all that time. In the 1980s David Attenborough put all of geological and human history on a time-line. He started with January 1 as the supposed creation moment, the Big Bang. In something like late April on this time-line, the first simple life forms crawl out of the oceans. In May and June lizards and salamanders and reptiles appear. By September the first primitive mammals arrive. Homo sapiens appears in the last three minutes of December 31. The last three minutes of December 31! And that's just Cro-Magnon man. The Hebrew-Christian tradition, which claims to speak with infallible truth, appeared in the *last second* of December 31.

Creation spirituality reveals our human arrogance, and maybe that's why we're afraid of it. Maybe that's why we are afraid to believe that God has spoken to us primarily in what is. Francis was basically a hermit. He lived in the middle of nature. And if we want nature to come to life for us, we have to live in the middle of it for a while. When we get away from the voices of human beings, then we really start hearing the voices of animals and trees. They start talking to us, as it were. And we start talking back. Foundational faith, I would call it, the grounding for personal and biblical faith.

But very quickly after Francis, we moved into monasteries, into buildings. We moved into cities, and even into universities. Our primary focus was no longer the land, the earth, the seasons, the sun and the moon and the stars.

I spent this past Lent in a hermitage. When we get rid of our watch and all the usual reference points, it is amazing how real and compelling light and darkness become. It's amazing how real animals become. And it's amazing how much we notice about what's happening in a tree each day. It's almost as if we weren't seeing it all before, and we wonder if we have ever seen at all. I don't think, brothers and sisters, that Western civilization realizes what a high price we pay for separating ourselves from the natural world. One of the prices is certainly a lack of a sort of natural contemplation, a natural seeing. My time in the hermitage re-situated me in God's universe, in God's providence and plan. I had a feeling of being realigned with what is. I *belonged* and was thereby saved! Think about it.

So, creation spirituality is, first of all, the natural spirituality of people who have learned how to see. I am beginning to think that much of institutional religion is rather useless if it is not grounded in natural seeing and nature religion. At the same time, I don't think we can make Francis into merely a nature mystic, either. He isn't. He clearly, again and again, praises the creatures *because of* the Creator. He praises God through the world. He looks at the stars and says, "If these are the creatures, what must their Creator be like?!"

In Francis, animals and nature are addressed not merely as animals and nature, but as spiritual beings who are part of the necessary harmony of things. It is almost as though Francis rediscovered within himself the universal harmony, the connectedness. He is part of a great chain of being. And he has no doubt that nonhumans are endowed with a kind of comprehension. He has no hesitation in talking to them—not in a cute or poetic way, but in a sort of sustained homiletic fashion.

We probably don't communicate with something unless we have already experienced its communications to us. I know by the third week I was talking to lizards on my porch in the hermitage, and I have no doubt that somehow some communion was happening. I don't know how to explain it beyond that. I was reattached, and they were reattached.

One day, I even asked the Lord to give me some sign that I wasn't just imagining this. I said, "Are they really looking at me? And am I really looking at them? Or, am I just making this all up?" And as I said that, a bird flew up. There was a glass door between me and the porch at the hermitage. The bird flew right up and hit his beak against the glass door. Then he hopped down and stared at me. I'm not going to say he came in—he didn't—but after pecking at the door, he hopped back a few feet, looked up at me, and then flew away. Accident? Coincident? Synchronicity? Illusion? Realignment? Believers simply speak of providence and miracles—and take them for granted.

When we are at peace, when we are not fighting it, when we are not fixing and controlling this world, when we are not filled with anger, all we can do is start loving and forgiving. Nothing else makes sense when we are alone with God. All we can do is let go; there's nothing worth holding on to, because there is nothing else we need. It is in that free space, I think, the realignment happens. That's the realignment out of which Francis lived. And I think it is the realignment that he announced to the world in the form of worship and adoration.

But let me give a few examples from Francis's life. There is his famous "Sermon to the Birds." Let me point out that when this was given, he has been living as a hermit, and he is still very attracted to going off to a life of prayer. He asks Clare and Sylvester to pray and to give him God's will for him. They come back and tell him that, yes, he can certainly take some time out in the woods and in the hermitage, but he is to be a preacher. And so, with great excitement and without wasting a moment, "he set out like a bolt of lightning," Celano says, "in great spiritual ardor, not paying any attention to road or path." As soon as he gets the word that he is supposed to be a preacher, he goes to preach! Well, he just can't wait, and it happens that the first thing he sees is a field of birds. So he realigns himself with the world; repeats and reconciles the universal harmony that he's discovered within himself; and speaks to them.

> My little bird sisters . . .
> You owe much to God, your Creator, and you must always and everywhere praise him. Because he has given you freedom to fly anywhere. . . . Also, he has given you a double and triple covering and your colorful and pretty clothing. And your food is ready for you without working for it. And your singing that was taught to you by the Creator. And your numbers that have been multiplied by the blessing of God. And because he preserved your species in Noah's ark, so that your race should not disappear from the Earth. And you're also indebted to him for the realm of the air which he has assigned to you. Moreover, you neither sow nor reap, yet God nourishes you. And he gives you the rivers and springs to drink from.

In most of his talking to the created world, Francis sees an interdependence between the different layers of creation, and he always praises them for what they give one another.

> He gives you the rivers and springs to drink from; he gives you high mountains and hills, rocks and crags as refuge, and lofty trees in which to make your nests. And although you do not know how to spin or sew, God gives you and your little ones the clothing which you need. So, the Creator loves you very much.

God loves these creatures. God has given them so many good things. Of course, the assumption is that if God loves them, they are objectively lovable; therefore, we are to love them.

> Therefore, my little bird sisters, be careful not to be ungrateful. But strive always to praise God.

They are to live in gratitude, which, of course, undoubtedly reflects Francis's own soul. We see also in the "Canticle of the Creatures" that Francis sees all the world as family. Everything becomes brother or sister. I think that comes out of a mystical insight, a contemplative insight, an insight that recognizes that we are a part of this great chain of being, that these are brothers and sisters, and therefore we may not disrespect them. This becomes the beginning of Francis's universal worldview: all created things are a mirror that reflect the Creator.

It seems that Jesus wished to emphasize, when he talked about the sparrows, that God cares for humanity the same way God cares for the sparrows. Francis, however, uses the phrases to stress instead the birds' status as a part of creation: "And God's special favor that they enjoy." It is a sign of the birds' prestige and position that they are taken care of without working. They have their own niche in the chain of being and their own special status before God.

How did Christianity move so far away from such a natural worldview, from a natural worldview that respected the earth? Those who are looking for a scripture quote to blame it all on usually go back to the first chapter of Genesis. It seems here that God is telling Adam and Eve that they are not to celebrate this creation but to *dominate* it (v. 28). That verse has been used century after century. It is amazing that we ignore so many other verses, yet get a lot of mileage out of this one! It has been quoted everywhere and *assumed* to be true. "Be masters of the fish of the sea, the birds of the heaven and all living creatures that move on earth." *All* of them, it says. That implies an anthropocentric world, with Homo Sapiens as the pinnacle of creation. All God really cares about is us; everything else is there to serve us. We can kill, use, exploit—because everything is there for us. Now, more than ever, especially after the horrendous 1980s, our arrogance appears to be catching up with us.

It used to take twenty thousand years for a single species to go into extinction. Now a species goes into extinction every twenty-five minutes. That is a sad result of the technological century in which we live. Domination has come to its high point, which is, in fact, its low point.

God said, "I give you every seed-bearing plant all over the Earth; and every tree that has seed-bearing fruit on it, to be your food." Many native peoples to this day pray and ask forgiveness of a tree they need to cut down. "Brother tree, I need you for this purpose. Will you forgive me for cutting you down? I will use you for a good purpose." Native American people had a strong creation spirituality, and I think it is still deep in the American psyche.

Older Franciscans will remember that a friar had to go to the provincial to get permission to cut down a tree. At least, that was still true thirty years ago. Franciscans did not cut down a tree with impunity. That survived. Isn't it amazing? Only the highest superior could give permission to cut down a tree. I don't know where that law came from; I doubt it went all the way straight back to St. Francis. But, certainly, we have lost that attitude of natural reverence, that awareness that God has allowed this tree to exist and therefore who am I to decide casually that it doesn't need to exist?

Our great thinker Blessed John Duns Scotus took this creation insight of Francis and made it into a philosophy. He taught a doctrine of "haecceity," the individuality of everything. God did not create genus and species, in Duns Scotus's thought. God only created *individuals*. Amazing! He took Francis's insight and moved it to the level of philosophy. Things are not created just as categories. We are one of six billion human beings, but God has created us in our "this-ness." And God, right *now*, is preserving us in our "this-ness." He's choosing us again, right now. Extraordinary! And, once we know that, once it's true of us, once it's true of the whole chain of being, we don't lightly step on worms. We don't lightly destroy anything, ever again.

The choices Francis made were made for very real socio-political-economic reasons. That is certainly true of his refusal to *possess* or to *own* or sometimes even to use. If we make an idol of possession, if we make a god of private property and possession, we soon accept a worldview of domination and control, which inevitably justify and feed on violence. Francis avoided the whole thing and said, "Don't possess." Nonpossession is the only way to be free from a life of domination and violence. Once we get into protecting our possessions, once we get into a worldview that idealizes possession, it is just a few steps before we have to increase and protect our possessions. Then we become part of a world of violence, power, and prestige.

I'd also like to quote from Francis's "Salute to the Virtues" to show how he understood this great chain of being—that we all are interdependent and rely on one another and owe respect to one another. This is the conclusion to his "Salute to the Virtues," and he's talking about obedience, of all things. Sometimes Francis almost sounds like a Buddhist: detachment from our own will, detachment from our own agenda, detachment from our own need to have anything. He says, "Holy obedience puts to shame all natural and selfish desire. It mortifies our lower nature, makes it obey the spirit, and makes us obey our fellow men and women. Obedience subjects a person to everyone on earth." So, there's a universal meaning to obedience, not just a vertical one, and that is *listening*. But then Francis says more, "And not only to men, but to all the beasts as well." Talk about submission! Here we have a man who is willing to say that he is not only going to submit to the will of other human beings but to the animals. "To the beasts as well and to all the wild animals, so that they can do as they like with me as far as God allows them." Here is a man who has turned our world upside down. He is not in domination over anyone. He is in such awe, such wonder, such respect, such obedience before creation that he gives it power over *him*. Extraordinary. There's a holy fool. The holy fool is always pictured in mythology as the court jester standing on his head. Why? Because he turns the world upside down. And if Francis is a holy fool, which he certainly is, he, like Jesus, is always turning reality upside down, always saying, "It's not what you think." The way the world is ordered, the way the system orders itself, the way the world of domination and power orders itself, is not the way *we* must order ourselves. This is the gospel's "new world order."

As a young man Francis went out into the backyard and looked at the stars. He was all excited about the stars. Francis was excited about everything—parties and fun, and friends—but he was sitting there in awe before the stars. And he said, "If these are the creatures, what must the Creator be like?" That is his starting point. If these are the creatures, what must the Creator be like? His first response to creation is to give praise for it, because it reflects the fingerprints of the Creator. He is first of all grateful for it, and therefore in love with it. This is probably why Francis is so universally popular: he begins not with a strong sense of original sin but with a strong sense of original blessing. Creation is, first of all, good, very good. He begins with the positive. He knows we cannot build on death; we can only build on life. So, Francis starts with the positive, saying, "What *is*, is first of all to be trusted, first of all to be loved, and first of all to be seen as a gift of God." He begins with the vertical, a world of relationships and reciprocity, of giving and receiving, of interdependence among the different levels of the chain of being.

We know of Francis's massive disillusionment with his father. He goes to the edges of the earth to create his family in order to redeem the pain of that first father relationship. And he calls this new one, "The family of those who serve God." That family becomes the Clares, the Friars, and finally, all of creation.

Many people see the world as hostile. This is apparent in the paranoia of the modern age. But do you ever find yourself, probably when you're alone, with a natural, trustful smile coming to your face while you're just looking at what is? Just looking at the flowers in your yard? That's probably the best giveaway that I know for our real belief in the world, our real operative image of the world. Does it give us delight? Are we able to let it give us delight? Or is it something we constantly have to protect ourselves from, that we have to back off from? It doesn't stay an indifferent universe for long. Without faith, the normal direction is toward mistrust, fears, anxiety.

If our worldview is basically indifferent, it won't be long before the hurts of life turn the world into a hostile universe. Then we are not at home here anymore. We can't be a contemplative at that level, because we basically are not able to see. We put up blocks between us and what is. The most common block that we put up is to go into our head and look for formulas and explanations and judgments. I have come to see why Jesus told us not to judge. Judgments don't have much to do with the search for God, with the search for truth. Judgments normally have to do with the search for control: that's right, that's wrong, that's good, that's bad, that person is virtuous, that person is sinful, that person's a Catholic, or she's a Protestant, that person is American, that person is a Mexican. There is some kind of sense of control here that gives the ego great comfort, because we "know where we stand." Then we try to surround ourselves with people who won't threaten us: other nice, white, middle-class Catholics. We won't learn very much this way. We won't become very wise, because basically we see the universe as hostile and are on guard. We cannot trust it, or entrust ourselves to it.

And entrusting ourselves to God? We can't. If we can't entrust ourselves to a black person, to a brown person, to a Protestant person, to an Iraqi person—to anybody outside our comfort zone— how can we possibly, brothers and sisters, entrust ourselves to the Wholly Other that we call God? When little "others" are already a threat to us, the world is a scary place filled with threatening people who aren't like us. They weren't educated the way we were educated. They don't think the way we think. They don't talk our way. We must exclude them from the table of fellowship, because the only people who sit down at the table of fellowship must be like us. That is always the temptation of the church, the temptation of every community, the temptation of history. It's an immense temptation right now, because there are five billion of us on this planet, and it's getting scary.

Twenty years ago we were talking about integration, and now we don't know if we want it anymore. We're pulling back into ethnicity, ethnic cleansing, getting into our own group and justifying why our group is right and safe. We judge that our group's going to heaven, and your group's going to hell. Franciscanism has to be something more than that. But it won't be, if we haven't come to the contemplative insight that Francis came to, if we don't come to know in the lowest level of our gut that *it's OK*. Despite all the problems. Despite what's happening in this diocese, in this religious community, in the church, with the papacy, in our own soul, it's still radically all right. Only grace can give us that, only radical grace. We can't convince ourselves. It is the work of God in the soul that tells us that despite all the pain, it is still a benevolent universe, it is God's history, God is in charge, God is the alpha and the omega in Jesus, and God is going to bring this together. *We* can let go.

If we *don't* know that, what else would we do, sisters and brothers, but spend our life trying to secure ourselves by storing up money, building our reputation, acquiring whatever else we think is going to make us happy and secure?

This is the bridge to Franciscan poverty, this is the bridge to Franciscan nonviolence. But it has to begin with this radical sense that all is well. That's the peace that we find in the saints. There has to be a place where we can give it back to God, and know God is still in charge. And we do not have to explain it, even to ourselves; we do not have to understand it. We must, as Francis did, merely "stand under" it. To stand under is probably the only way to understand it. And this is to enter into the realm of mystery, where creation becomes, I think, our primary teacher. What is. The Bible of the world, the Bible of nature, the Bible of the sun and the moon and the stars, the Bible of the seasons, the Bible of light and darkness—it's all there for us once we meet reality in its most simple and direct form.

Go through the woods. We see young trees, and we see trees in their great height and health, and we see dying trees. And somehow, our spirit says, "I was once a little sapling, but I'm going to be one of these old trees pretty soon, and that's OK and even good. I'm part of the Great Wheel, and it's part of the great mystery, and every day is darkness and light, life and death." This is the Paschal Mystery that Jesus walked and trusted.

I was preaching in California, and I told them, "It's dangerous to live in California, because you don't have the truth that they have in Minnesota. Minnesota speaks reality better than Southern California. So don't stay here too long; take a few trips up to Minnesota." Half of life is cold; half of life is scary and dangerous. Only half is sunny summertime. Together they make a full cycle. That's the inexorable wheel, the wheel that we Christians call the Paschal Mystery. It is also good creation spirituality. It's as solid as the scriptures, it's as solid as the psalms, it's as solid as St. Francis. Many people are attracted to creation spirituality, which is really Franciscan spirituality, because they recognize the need for again learning how to see. Once we learn how to see, how to contemplate, how to let the moon and the sun and the stars and the seasons and the trees and the plants and the animals speak to us; once we find our instinctual self in the eyes of animals, then we are ready to go and study theology.

The trouble with some of us is that we studied theology *before* we learned to ride the wheel and feel our instincts. The temptation is to go up into our head and stay there, looking for control, looking for formulas, trying to understand, eating from the tree of the *knowledge* of good and evil instead of walking into the rest of the Bible. Unless there is at least some ability to surrender to mystery, to see the darkness and light, and the inexorable and real unfolding of God in history, I think that an awful lot of us are going to become bitter and angry people in the next ten years, because I don't see it getting easier. I think the next ten years in the American church are going to be very hard. Creation spirituality might well serve us in this time.

Franciscan Nonviolence

I believe the church has grown a year a century in its capacity to comprehend Christ. This means that five hundred years ago we were just in the first years of adolescence, as it were. We were in what's called the heroic period of our lives. We usually think of the word *heroic* as a positive word, and it is, but the heroic period also means that period when we are trying to define ourselves as great, when we are trying to prove ourselves, assert ourselves, name ourselves, identify ourselves.

When we are named by God, we don't have to self-name. When we are secured by God, we don't have to self-secure. But we are not radically in the arms of God. We spend much of the first third of our lives self-securing, self-identifying, self-naming, trying to give ourselves a face and a form to prove that we're important and significant and good and strong and powerful. This is particularly true of the male of the species.

And because it was still very much the patriarchal period of our history, the males of our species, the Conquistadors, pretty much defined the gospel. They defined it not in the terms that Jesus defined it, powerlessness, but in terms of where the male psyche is in its teenage years, power. If I had to name the one sin, the one blindness, the dark side of the Roman Catholic Church, it would be its relationship with power, its inability to recognize what Jesus clearly said about powerlessness. That continues to this moment, to this day. *It is, in my opinion, the sin of Western Christianity.* So, if we didn't understand Francis, it's because we didn't first of all understand Jesus. And, of course, if we're so easily able to dismiss what Jesus clearly says about nonviolence and powerlessness, it's no wonder that we couldn't see it in Francis.

We might see the beginning of Francis's massive loss of innocence, his loss of idealism, the entry point to his great dark night, if you will, as being after 1219 when he returned from the Sultan.

He has about seven more years to live. And that certainly is Francis's great dark night of the soul, the last seven years of his life. He went to announce to the Sultan the way of peace. He came back seemingly having made friends with him to some degree, but certainly with no tangible results.

Now, brothers and sisters, we need to place this in historical context, and the easiest way to do that is to compare the situation with the recent Persian Gulf War. Picture the yellow ribbons everywhere, and everybody jumping up and down with American flags, cheering for our troops to kill Iraqi people in the name of our Christian country. What if one of our bishops, or heads of state, made a flight to visit Saddam Hussein? What if you, at your own expense and trouble, did so? Do you think you would have been appreciated, even by your own religious community? Some convents were filled with yellow ribbons and American flags too.

We have to realize the revolutionary, countercultural character of Francis's response to the Crusades. He went over there and tried to talk to the Sultan personally. Extraordinary! When he returned he realized he had not had any great practical success. That's one of the great and *necessary* turning points in our life with the gospel; we have to confront the terrible recognition that the gospel is not about success.

We are a culture that has for some four- or five-hundred years been based on a philosophy of so-called progress. But the psychology of growth is only the psychology of the first part of life, usually, the first fifteen or twenty years. The psychology, and therefore the spirituality, in the rest of life is not the psychology of growth. It's the psychology of diminution, of letting go. There's something very dangerous about a language of endless growth and progress. It's wonderful on some levels, but it is too one-sided, and it falsely defines wholeness. The gospel names reality as the Paschal Mystery—the inexorable wheel, the dying and the rising of things.

The bishops have put out a document on the Quincentenary. I have to admire them; they were very honest. In the document they say, "The church, with few exceptions, accompanied and legitimized"—accompanied and legitimized doesn't say we did it, we weren't in the front line, we were just blessing it from the second line—"the conquest and exploitation of the New World." The *zeitgeist*, the spirit of the age was too strong to stand against. Forcing conversion to European forms of Christianity demanded a submission from the newly converted people that facilitated their total conquest and finally their total exploitation. There's no point in imputing blame or shaming anybody at this point. We can't judge the past from our vantage point. But we've got to be truthful about what we didn't see, so we can see better now.

Let's look at Columbus's diary, the second voyage in 1495. He writes, "They do not bear arms, and they do not know arms. They would make fine servants. With fifty men, we could subjugate them all and make them do whatever we want." Make them do whatever we want. Now, why these people would become Christians after that is beyond me. But, by the grace of God, many of them did. Hopefully, they have experienced also the freedom and the good news of the gospel that was brought to them. Maybe that's the only way the gospel ever gets through. The French were all baptized by coercion, as well; this didn't start with the Indians. We moved in and poured water over the French, and over the Germans, and over the Spanish. Remember, that's the way adolescents do things—power.

That's the part of a heroic consciousness; we win by conquest, by power. The human psyche, the human consciousness, probably wasn't capable of much more than that. That's why Francis was such an extraordinary person in the thirteenth century. He seems to be a man literally born out of time. We can see why he is called an *Alter Christus*.

Sisters and brothers, the discovery that's come to us, and now it seems so obvious that we wonder how we failed to see it, is that the history we have, the only history that lasts, the only history that passes down to us is from a very small group. History is written by those who have the time and resources to get an education. Which means that history is written by the educated and the wealthy. Narrower yet, it was normally men who got an education. And even among men, it was very often only the clergy!

I hope you understand the immense and extraordinary decision Francis made when he called us "O.F.M."—Brothers of the Lower Class—and what socio-political issues he was dealing with. He was trying to get us to change classes, to stop this exclusive interpreting of the gospel from the elite side. To read the gospel with the eyes of security is to misread it. To read the gospel through the eyes of the monied class, the upper class, those in charge, those who name the questions and themselves give the answers, is to hear self-securing answers, not the dangerous voice of Jesus.

The Bible is one piece of enduring literature that was not written by the conquering class, but rather by the losers. Ninety-five percent of the Bible was written by people who were oppressed or occupied or enslaved or poor. That's extraordinary! That kind of literature doesn't last; it never even gets written. And that's what makes the Bible such an extraordinary book. The only real exceptions are the books of Leviticus and Numbers, which were written by the priestly class, and the books of Chronicles. And those books bore us to death. They're all preoccupied with the sanctuary, the sanctuary, the sanctuary, the colors of the candlesticks. Does this sound like anything recent? Hear Jeremiah say, "Stop talking about the Temple, the Temple, the Temple. What's happening out in the streets? What's happening out in the villages? What's happening among the people?" (Jer 7). Jesus also said, "The Temple has to go. Not a stone is going to stand on a stone." Jesus said that right after the disciples had been praising the fine architecture and building materials of the temple (Mk 13:1)!

At St. John Lateran in Rome, there are wonderful, heroic statues of the twelve apostles holding up the church, great heroic figures, James and John and Andrew and the rest. Actually, we know hardly anything about them; the only thing we do know from the gospel is that they misunderstood Jesus until the very end. Where did this myth come from, that they held up the church, created the church? Maybe they did, but there's no strong scriptural evidence for it. Mostly they fought Jesus all the way. In the last chapter of Mark, they're still afraid. Yet we create this myth that they created the church. Now, I'm not down on the twelve apostles, although it sounds as though I am. All I'm saying is that, from the text, there's no real evidence that they got the point. Peter is presented, certainly in the synoptic gospels, as a buffoon, a buffoon who misses the point consistently and constantly. Yet Jesus loves him and forgives him. Peter is certainly not presented as a power figure. He just loved Jesus immensely, and I presume that transformed him to be like Jesus—nonviolent and crucified upside down.

Could that be a metaphor for the whole pattern? I think so. Could it be a naming of how God is willing to deal with the church and willing to deal with history? Romano Guardini said in the 1950s that "the church has always been the cross that Christ was crucified on." Doesn't that hurt? And yet, maybe it's true. The church has always been the cross that Christ is crucified on. The church that does not believe its own gospel.

Let's go through a little bit of history to see how we got to this point. Let's start with Jesus himself, and hear some of his rather clear and self-evident teaching of nonviolence. "You have heard how it was said: *Eye for eye and tooth for tooth.* But I say this to you: offer no resistance to the wicked" (Mt 5:39). Extraordinary. How do we even understand that? How do we deal with that in a heroic culture? How do we deal with that in a culture of Conquistadors?

If anyone hits you on the right cheek, offer him the other as well. If someone wishes to go to law with you to get your tunic, let him have your cloak as well. And if anyone requires you to go one mile, go two miles with him. Give to anyone who asks you, and if anyone wants to borrow, do not turn away.

You have heard how it was said, *You will love your neighbor and hate your enemy*, but I say this to you, love your enemy. (Mt 5:39-43)

This is the most radical teaching on this earth. "Love your enemies." How can we possibly do that? "Pray for those who persecute you, so that you may be children of your Father in heaven, for he causes his sun to rise on the bad as well as the good, and sends down rain to fall on the upright and the wicked alike. For if you love those who love you, what reward will you get? Do not even the tax collectors do as much?" (Mt 12:44-46). Big deal, he's saying. If we greet only our brothers, what's so great about that? We are not enlightened if we can only mix with clones of ourselves. What's great about that? There is no liberation evident at all if we just love those who love us. "Do not even the gentiles do as much?" Jesus asks. "You must therefore set no bounds to your love, just as your heavenly Father sets none to his" (Mt 5:48).

Jesus' teachings, and most of all his own example on the cross, present a seemingly impossible faith ideal, one which cannot be understood outside the realm of grace and faith. I think that over two thousand years only a very small portion of believers have heard this gospel. It emerges here and there, normally to the degree that we have *not* bought into the system. If we are getting our rewards from the system, if we are getting our job security from the system (I think this is why Francis wasn't ordained a priest), it's very hard to speak truth to that system, because, frankly, our job is at risk. Maybe that's why Jesus told the apostles to leave their nets. He had to detach them from job security. I wonder if Jesus even intended professional religious leaders, when I see what we've done to the gospel century after century. We've courted power and money and prestige and riches, and not just riches but opulence.

So, how did we get from Jesus of Nazareth to palatial luxury, while saying we were infallibly following Jesus? We can take what Jesus taught, and yet can practice exactly the opposite, because most of us can't criticize what we ourselves desire—power, prestige, and possessions. (I still wonder why we priests walk around the sanctuaries with tassels at the bottom of our stoles. You know, you'd think we'd go out of our way to never have a tassel at the end of a stole. After all, Jesus literally makes fun of this [Mt 23:5-7]. Yet we've got tassels all over the place. Now, I'm just using that as a silly example, but in general, it's true.)

Jesus' clearest, most self-evident teaching is a teaching that we have consistently ignored. Jesus' teaching is clearly and provably concerned with power, prestige, and possessions—and their protector, *violence*. His words to Peter are very specific and literal: "Put your sword back, for all who draw the sword will die by the sword" (Mt 26:53). Wouldn't you think if violence is ever justified, it would be defending the incarnate Son of God? If Jesus isn't teaching nonviolence, what is he teaching? No one is really devious in denying his teaching. He is just too much for us, and we have not been ready to hear him.

There is a common avoidance technique at work in most religions, and the best way to avoid the message is to get preoccupied with worshiping the messenger. But we are ready for the message now. Human consciousness and Christian consciousness are secure enough, our boundaries are secure enough, our identity is secure enough. We saw this at Vatican II. The Catholic church finally grew up enough that it could stop defending itself. It probably takes people at least twenty years to attain the security to stop showing that we are right and other people are wrong. We move into this art in the second half of life. We start to study the art of letting go instead of the art of self-development and self-growth. Adults alone are ready for the folly of the cross, the risk of nonviolence, the vulnerable message of Jesus.

The message holds on after the death of Jesus. Paul, for example, says, "Bless your persecutors; never curse them, bless them. Rejoice with others when they rejoice, and be sad with those in sorrow. Give the same consideration to all others alike. Pay no regard to social standing, but meet humble people on their own terms" (Rom 12:15-16). It seems we've got to break our identification with the empowered class. "Never pay back evil with evil, but *bear in mind the ideals that all regard with respect. As much as is* possible, and to the utmost of your ability, be at peace with everyone. Never try to get revenge: Leave that, my dear friends, to the Retribution. As scripture says: Vengeance is mine—I will pay them back" (Rom 12:17-19).

But there is still confusion among good Christian people. On the evening news, for example, we hear that someone was killed and the parents want justice. They don't want justice; they usually want vengeance. We have not taught people how to distinguish between justice and vengeance. An eye for an eye and a tooth for a tooth is pagan. And putting people in jail or killing them, as good Catholics are wont to do, does not achieve a bit of justice as such.

But there is more. "*If your enemy is hungry, give him something to eat; if thirsty, something to drink. . . .* Do not be mastered by evil, but master evil with good" (Rom 12:20). In fact, Romans 12 could be used, if you take it verse by verse, as a magnificent, full teaching on nonviolence. Nonviolence does not mean turn over and play dead. Nonviolence does not mean that Christians are not to resist evil. It means that we have to be a lot more creative and imaginative and faith-filled in the ways that we resist evil. We do not, in the first instance, jump to the worst alternative, "Kill them!" That is not an acceptable Christian response.

It's simply clear that Christians do not kill, period. Once we create the moral foundation, then the moral foundation has to be applied to abortion, to capital punishment, to euthanasia, to war; it has to be applied, finally, to the killing of the earth. Christians don't kill.

Of course, liberals choose one side and conservatives choose another. Both are inconsistent. If we have come to the moral conclusion that killing is immoral, then killing is immoral, period. God alone has choices over life and death. That's what Jesus and Paul are talking about, but it has taken us two thousand years even to *begin* to admit that it might be true.

The Smithsonian Institute in Washington says that the longest bibliography of any human person in history is that of Francis of Assisi. He's got the longest listing of biographies, monographs, periodicals, and studies of any person who ever lived on this planet, even more than Jesus. Francis! This man is larger than life. Culture after culture, age after age, group after group is enamored with the extraordinary life of this man. How could he break through and take Jesus so literally in precisely the fundamental issues of the gospel? (What we have in contemporary fundamentalism, however, is a very mechanical interpretation of peripheral passages while ignoring "the greater matters of the law" [Mt 23:23].)

Lower-class Solidarity

I mentioned before that Francis called us the Order of Friars Minor. In twentieth-century sociological language, we would call ourselves the Brothers of the Lower Class. *Frati Minori.* Francis meant a change of social position. We all know that. We were all taught that in our first studies of Franciscanism. Francis was from the middle class, the new mercantile class in Assisi, and he clearly changed sides from the *majores* to the *minores*.

Conversion, metanoia, "turning around" is essential if we are to be honest. We must recognize the agenda that we are protecting. What we see in Francis is clearly a changing of class and an identification with a new viewpoint; an identification with the side of powerlessness. We have a fancy phrase for that today—the preferential option for the poor. In the last fifteen years that phrase has been adopted by the constitutions of 86 percent of the religious communities all over the world. Now, that's the *sensus fidelium*, right? That's the church recognizing the rediscovery of the gospel. We *do* have to be biased. We have to make a choice.

Jesus ate with the rich. But he never sat with the rich without challenging them. He doesn't shame them, he doesn't say, "You're lost and going to hell," or anything like that. He simply points out that the system makes it hard to know the truth, hard to ask the real questions. To get at the truth inside of any system, ask the people at the bottom, not the people at the top. In Jesus' words, "The last will be first and the first will be last." He says, "You must invite the little ones to the table—the lepers, and the blind and the lame, and the drunkards and the tax collectors." All those we leave outside have a gift of truth for us. "The wounded one is the gift-giver" is a description of the phrase "Jesus saves."

The church never has believed this. In my opinion, most of the church has been in clear heresy on this since the beginning, because it does not believe this clear teaching of Jesus that the truth is at the bottom instead of the top. We have glamorized the top, and in that we have imitated the world. Jesus turned reality upside down in his announcement of the kingdom. He said those at the bottom have a better chance of knowing where the pain is, where the lie is, and therefore, where the way is. We know that is true, especially any of us who work with the homeless, with refugees, with those who are disadvantaged or oppressed or victimized or marginalized. They know what is really happening in the system. As long as we listen only to white, rich males, we're never going to get very far, because, rich, white males keep redefining the world in terms of their own system of success. That's the culturally tragic position we're in today. Because we've continued to look to white, rich males to tell us what the truth is, we've probably reached an all-time low point, a decade of denial. We deny pain and the oppressive nature of the system. Now we have poverty out of control in this country. Do you realize the word *homeless* was not even in the dictionary in 1980?

Francis and Jesus found a way to stand against that inevitable movement of culture, and their agenda was to change sides. Change sides. That's the privileged position of the gospel, so much so that those of us born in the middle class, particularly white, male clerics, educated male clerics, have an obligation to change sides. That's what Francis believed.

If you want to read some exciting studies on this, Arnaldo Fortini, the mayor of Assisi, wrote a good biography. He was the first biographer who had access to all of the historical and economic records from the time that Francis was alive. So, his life tends to be a lot more honest. For example, he finds out (and this explains some of the tension between Francis and his father), that Francis's father was not only of this first generation of new tradesmen, but from the immense money he made, he was buying up the little farms that were going under around Assisi. He was a landlord who was oppressing the poor family farms. That explains, I hope, some of the very real rage and radicalism of Francis's poverty. He had no respect for his father. He saw his father taking advantage of other people's disadvantage. Francis refused to be a part of that profit. This shows, I might add, that Francis had good social analysis skills. He wasn't naive and apolitical. He saw how people were made poor and were kept poor by the system. He knew that a person doesn't become rich without stepping on a lot of folks, without taking advantage of a lot of loopholes. We don't get to the top of any system by powerlessness, by vulnerability, by the way of the gospel. As Merton says, "You spend your whole life climbing the ladder, and you find it's against the wrong wall."

There's been no real metanoia, no conversion until we move from the side of privilege and power and prestige. Jesus does not say the big problem is sexuality; Jesus does not say the big problem is obedience to Peter; Jesus does not say the big problem is people skipping Mass on Sunday. In the Sermon on the Mount Jesus says the problem is power, prestige, and possessions. And the church has been in bed with power, prestige, and possessions for centuries! I say that not as a disobedient or disrespectful son of the church, but as a son of the church who loves the gospel.

We've lost moral credibility in the West. And if we are not morally credible, it's because we have not believed our own founder. We've not taken his teachings seriously. We offer lip service to Jesus, but we don't care a bit about what he really says. We've been on the wrong side of most of the revolution for the last five hundred years— always on the side of the Batistas and the Somozas and the wealthy and the powerful.

The pope is absolutely right in telling priests not to get too deeply into the political system—although he does himself—because once we are in it, we have to buy it, to a great degree. The best position is to stand outside, and then we can keep speaking the truth, and living the truth, and being the truth, and naming the truth. We have no debts to pay to anyone except to God.

On Poverty

I don't think we can keep interpreting the vow of poverty simply in terms of a private thing we do, which is supposedly pleasing to God. We need to see our lives, especially our lives in common, as opting for a new economic order, a new way of living based much closer to the communist end of the spectrum than the capitalist. In his 1981 encyclical *Laborem Exercens* Pope John Paul II recommends "modified socialism." Did you know that? His statement got almost no press in America. We could not believe that he could possibly, conceivably, say such a thing, so we didn't hear it. We just pretended he didn't say it.

There have been many things the pope has said in terms of social justice that we just can't hear, because he criticized countries like America. He recognizes that we have been just as much of the international problem as the communist countries of the world.

What the pope recognizes is that we were meant to choose the low road, to enter into solidarity not with the upper class but with the lower class. How would the evangelization of the Americas been different if the church in general, and the Franciscans in particular, had seriously identified with the poor and taken the side of the powerless?

No one questions that the early Franciscans who came to New Mexico lived very simple, poor lives—lives of tremendous self-sacrifice. But the vow was still interpreted in terms of individual spirituality, not as an identification with the lower class. What if they had identified with the Indians instead of with the Spanish?

Look at the example of Our Lady of Guadalupe. What does Mary do, in Mexico City? She appears to an Indian, and she doesn't speak Spanish. Rather, she speaks his language. Mary changes sides! The Spanish think they brought Mary over, and what does Mary do? She changes sides. Mary understands the preferential option for the poor, because that's the gospel. In every case Jesus identified with the oppressed group. Whose side does Jesus take in a war? Jesus is on both sides of every war, because he is always on the side of suffering, on the side of the pain. There is no way to use Jesus for one side of any war, although we all try to. He changes sides if we are on the side of power and domination.

When Jesus' disciples tried to take the way of righteousness, Jesus changed sides against his own! (see Mk 9:38-40). Whenever they tried to take control of the system, the moral high ground, he changed sides. He's extraordinary. When they tried to take the side of male righteousness and power, he took the side of the women. He broke the rules; in fact, he just ignored the rules, as when he talked to the woman at the well. Or when he invited Mary to come into the living room with the men and talk theology. Women stayed in the kitchen and did women's work, yet Mary sat with the men. When Martha tried to get her to come back and take her "rightful place," Jesus said, "No, I'm not going to make her do it."

How did he come to this kind of freedom? I think the insight we must grasp is that Jesus always takes the side of pain, always takes the side of powerlessness, always takes the side of the poor, always takes the side of the victim, always takes the low road. It finally leads him to the cross. When we stand with the poor, we are not in the system. We are not appreciated by the powers that be, sometimes even by the powers that be in the church. The way of the powerless is inevitably the way of the cross.

Those examples where people are enslaved, where people are abused, come from our inability to live the gospel. We fail to recognize that Jesus is talking about a new world order, not just the private salvation of individual souls. The gospel is about the salvation of nations, which is exactly what the end of the gospels say: "Go preach the gospel to all the *nations.*"

Catholicism *at its best* has been very good at a social reading of the gospels. It has read the gospel in terms of a corporate message and not just an individual message. The gospel is fully and rightly read out of a context of society—what's good for the nation-state, what's good for the people, what's good for the village. We still see this in European architecture, which always has the plaza at the center. All the buildings open onto a common space, because the common space is the heart of the city. The common space is the heart of who we are, but we lost that sense of existing primarily for the common space.

Since the beginning of civilized history, since we've been able to write about it, 97 to 98 percent of the people who have lived on this earth have been poor. When Jesus was preaching, the majority of people he was preaching to were coming from that disadvantaged position, that oppressed position, that outcast position, and so they immediately understood the gospel. The disadvantage we have, and this has never existed before in history, is that we live in a country with the relatively new phenomenon called the middle class. We are the first products of it, and most of us, even people who are fabulously wealthy, call ourselves middle class in America. Even people who are very poor still want to say, "I'm middle class." But what characterizes middle class or "bourgeois" thinking is that its members have been given just enough of the advantages of the system to think like the rich! We tend to think like the millionaires! We're given just enough comfort to numb us to the right questions—and the right answers.

When you have this phenomenon of a massive middle class, its values tend to be rather naive and simplistic. The Romans called this "bread and circuses"; people have just enough worldly comforts so that they stop thinking. Those who are kicked around, oppressed by the system, search for justice and truth. They ask,

"Why am I at the bottom of the system? Why, no matter how hard I work, do I always get kicked to the bottom? Why does the system never work in my favor?" They have to remain politically astute and recognize their viewpoint and the viewpoint of those on top. But those of us in the middle class have the luxury of not caring. We have a bed, a car, three square meals a day. We're basically comfortable. We have been bought off by bread and circuses; we're entertained to death.

Faith does not have to do with success or progress, in a worldly definition. Faith has to do with love—how love is able to move, and how love is able to be increased on this earth, and how love is able to happen on this earth. The only doorway that faith opens is the doorway to love, not the doorway to power, righteousness, control, prestige, or position. Faith leads us to a powerless place, to an empty place, to what Merton calls "the place of nowhere," an empty place where we know all we have is God.

I think that now, eight hundred years after Francis, we are finally able and ready to see. In fact, we must see God's new social order if we are to have any credibility for Franciscanism in the next century. This is what the world expects of us.

"No Baggage" Evangelization

We have a hard time understanding Francis because we, as a Christian church, have had a hard time understanding Jesus. If we don't face the issue with Jesus' teaching first of all, we are inevitably going to mystify and spiritualize Francis. This is certainly true in terms of the Rule of St. Francis, which is, in great part, Francis quoting the gospels of Jesus. We must begin with the teaching of Jesus, and then recognize that Francis is a true religious genius in discovering Jesus' intent.

Jesus called the twelve apostles, and then he gave them their instructions. He sent these men on missions, as the twelve, and he gave them the following instructions: "Do not make your way to gentile territory, and do not enter any Samaritan town; go instead to the lost sheep of the House of Israel" (Mt 10:9). Now, let me stop right there. There is always a tension, in the work of religion, between exclusivity and inclusivity. The conservative type overemphasizes exclusivity; the liberal type overemphasizes inclusivity. A person has to be a real artist, a real faith-filled person, to know how to hold them in creative tension. Jesus does, very well. In this particular line, he's emphasizing exclusivity. In other words, a group has to have a sense of identity and boundaries before it can call people to it. We have to know who we are. We have to name ourselves. We have to walk a journey together for a while, and then we can say to the outsider, "Come, this is who we are." There's nothing wrong with boundaries. In terms of social psychology, they are necessary. We have to have boundaries, and at those boundaries we put a gate. We say, "This is what it means to be in, and this is what it means to be out." To do so doesn't mean we have to be exclusive; it simply means we are named.

Jesus is talking about the kingdom, which is beyond questions of exclusivity. "Bring in the lepers, bring in the blind, bring in the poor to the banquet." Jesus knew who he was. Only when we know who we are, however, are we ready to move out. So, the first thing he tells them is to find out who they are, what their group is. Once they have a sense of community, he can send them on mission. He says,

> As you go, proclaim that the kingdom of heaven is close at hand. Cure the sick, raise the dead, cleanse those suffering from virulent skin-diseases, drive out devils. You received without charge, give without charge. Provide yourselves with no gold or silver, not even with coppers for your purses, with no haversack for the journey or a spare tunic or footwear or a staff, for the laborer deserves his keep. (Mt 10:7-10)

This is clearly an image of what we would now call mobile ministry. Francis, of course, took this literally. But Jesus never tells his initial twelve apostles to set up a foundational, grounded place where people will come *to them*. That's the first stage of institutionalization, and Jesus actually gives us no foundation for institutionalization. It's clear that he is talking much more literally about a *movement*, and he uses every mythic image he can to keep us on the move, to keep us from institutionalization. Jesus talks about a movement called the Kingdom of God, and Paul turns it into the church. Maybe that's a little oversimplified, but there's a lot of truth to it. But we've got to be honest with the teaching of Jesus. At this point, at least, he is not talking about local churches and people coming to us. He's saying that we should keep going out to them.

Then Jesus says, "Whatever town or village you go into, seek out someone worthy and stay with him until you leave" (Mt 10:11). I think probably one of the greatest breakthroughs I made was in 1977 when I started giving retreats to third-world missionaries. The advantage of giving retreats to third-world missionaries is that you don't stay in American hotels when you go to those countries, but you stay in the houses of the missionaries, who are invariably living right in the middle of the barrio or favela with the people.

Normally, they would tell me to come a few days ahead of the retreat so they could take me through the neighborhoods of the people. And, if anything converted me, it was that. I had to come into their world, on their terms, eat their food. I couldn't even speak the language. That might be what it means to come "without baggage"— I think it is!

"As you enter his house, salute it" (Mt 10:12). Let the grace flow through you. Give your blessing to everybody. We call it the gift of magnanimity. Francis would have called it the virtue of courtesy or chivalry; he was always a knight. To bless others means to trust them, to empower them, to tell them, "You're doing it right, trust yourselves. You've got the power, you don't need to rely upon me for the power, the power is within you. You have been given the gift of the Holy Spirit, which has been poured out with generosity."

That's what John the Baptist was doing when he went to the River Jordan. He was saying, "God is as available as water in the desert." A person has to search for water in the desert, but it's there, and John was magnanimously pouring it over people's heads "for the forgiveness of sin." "You mean, we don't need to jump through hoops? We don't need to buy some turtledoves and sacrifice them in the Temple?" And John was saying, "No, God forgives us as readily as this. God is on your side more than you are on your own side. Your job is simply to announce the gift, to give the gift of the unconditional and free love of God, to keep telling the children of God that they are children of God."

Why is that so hard for us to do? All of these sacraments and symbols, they're all just metaphors, metaphors! They're all just fingers pointing to the moon. But the moon was there before we started pointing to it. We know that, don't we? Don't we?

In creation spirituality God has been speaking the truth since the beginning of time. We aren't inaugurating this truth. We are not initiating it. We simply are announcing this truth that always was. God has been loving us all through geological history. We're just the lucky ones who have come along now in a moment of time to bring it to consciousness, to give a word to it: Jesus. Jesus is the revelation of the heart of the Father. Jesus is revealing what's always been going on inside of God—and is going on inside of God toward Islamic people, Jewish people, Buddhist people, Hindu people. We don't have any corner on the market of God. We have a beautiful and clear and wonderful name for God. It's not going to get much clearer than Jesus.

We don't ever have to apologize for an *exclusive* relationship to Jesus Christ. Why? Because Jesus teaches total inclusivity! We don't ever have to apologize for an exclusive relationship to Jesus as our savior, lord, teacher, because when we draw close to him, we know that he is going to set before us a universal table. He invites drunkards and tax collectors, prostitutes, the lame and the blind, and Protestants and pagans too. Jesus preaches universal table fellowship. He preaches inclusivity with all of God's children.

Francis was proud of his Italian identity, his Catholic identity. But where did it drive him? To the Islamic people. Where did it drive him? To go visit the Sultan. Where did it drive him? To announce peace and good to the little people on the road. He didn't ask whether they were practicing Roman Catholics. A lot of them were Waldensians and Cathars and all those other heretics that they had in his time. His job was simply to say, "Good morning, good people! Do you know that you are good people? Do you know that you are sons and daughters of God?"

When will religion learn? It seems that in two thousand years, we should be getting the point by now. Jesus did not come to create enclaves of righteousness. The gospel is not intended to create in-house groups that can feel morally superior to other people. It's meant to create free people who have an excess of love. That's how we recognize people filled with the spirit: magnanimity and excess. If we have to hold on to it and protect it and define it and defend it, it isn't grace, and it isn't God. God does not need our protection. God just needs instruments and heralds. So Francis called himself the herald of the Great King.

I faced cancer last year. I was given two to six months to live, and I wish everybody could have that experience somewhere in the middle of life. It teaches us to go for broke. After facing death we have nothing to lose. And I believe this is the gospel. As Richard McBrien says, "I could probably put in a phone booth the people who have left the church because of Hans Küng. But I could circle the globe with the people who have left the church because of arrogance and righteousness." I could put in a phone booth the people who have left the church because of Leonardo Boff, and I could circle the globe with the people who have left the church because of the arrogance of the institutional church, because of the arrogance of Rome, because of my arrogance. We try to hold onto the grace of God and dish it out only when we think other people are worthy, only when we think other people are law-abiding and have played the game our way.

Brothers and sisters, the pain is too great in the world. The pain is too great to be niggardly with the grace of God. The discouragement is too great in the church, right now, to waste time criticizing the church; we don't need more criticism or negativity. We have the power to trust our own experience and to trust our own journey and especially to trust our own heart. It is in our heart that truth resides and grace is available and God is offered. God hasn't let us live our years without giving us a lot of human experience. We have to know, at this point, what works and what is real and what is alive and what matters and what lasts. Live it, love it, give it away with joy and abandon. There are too many people starving on this earth and too many people spiritually starving in the church.

When we announce the gospel with our lives, it appears to people like naivete. It appears that we are simple and pious and out of it. So, it's very easy for the world to marginalize or discount us. But we can trust our own journey, because God does. I've got to believe that. God lets us go down all kinds of dead ends. God lets us make all kinds of mistakes. Still, God has such authority that God can pull it all back together. We have to trust our experience; it's all we have.

The reason I can say that is because we are people who have thrown our experience together into community. We've thrown our experience together into the great mystery of the church. And so, I can say, "Trust your experience; trust your heart." To those of you who are women, I tell you that human consciousness finally has reached the point of honoring the feminine. But we desperately need, in this church, the recognition of feminine consciousness. It is the only thing that is going to, humanly speaking, balance out this church and lead us to the next stage.

I say to women: Trust your womanly hearts. Trust your womanly experience. Trust your feminine souls and offer them to your brothers; offer them to us. It will be easy, in the next years, to give up on us. It will be easy to say the patriarchal system is all corrupt and not worth giving any energy to. But, we are in this human thing together, aren't we? And God has made us male and female. And you're half of the truth and, for some unbelievable reason, I am, too. And that means that we've got to listen to one another, and we've got to hear the spirit of God in one another—male and female, rich and poor, American and non-American, Catholic and Protestant. Until we can hear God at that level, I don't believe we've heard God at all. I believe we've whittled God down to a little tribal god instead of the great God of history, the great God that Francis loved, the great God before whom he could say, "*Deus meus et Omnia*" ("My God and all things"), which of course is our motto. That's the God we all came to serve, and that's the God to whom we want to, again, give our lives. That God has not abandoned the church, that God has not abandoned religious life, and that God has not abandoned human history.

11

Fruit of the Earth, Fruit of the Vine

CHARLES CUMMINGS, O.C.S.O.

The sacrament of the eucharist, which contains the body and blood of the Lord Jesus in the form of elements taken from the earth, is the wellspring of the Christian community. Everything in the life of the church revolves around this sacrament, because it contains the vivifying presence of our Savior. Over the centuries the church has grown in its self-understanding by reflecting on the mystery of the eucharist. The dynamism of the eucharist progressively leads the church to its fullness and perfection.

Thus, in the post-Vatican II evolution of the Roman Catholic liturgy, we came to realize that celebrating eucharist had implications for social justice. Although the eucharist is the moment of our most intimate communion with the Savior, it does not support an individualized, privatized spirituality. The eucharist compels us to reach out to the least and poorest of all, to share our bread with them, even to lay down our life for them as Jesus did (Jn 15:13).

In the current era of ecological awareness, we are beginning to see that celebrating eucharist implies a commitment not only to our neighbor but also to all God's creation, the material world that is a reflection of "the invisible existence of God and his everlasting power" (Rom 1:20). As our neighbor is an image of God, so also is the material world. We are beginning to acknowledge that all creation is an extension of the neighbor whom we are asked to love as ourselves (Lk 10:27). The inner logic of the eucharistic sign of bread and wine is leading us to this new awareness. Our eucharistic vision is crystallizing.

Incarnation and Eucharist

Because of the incarnation, the chasm between the human and the divine has been bridged. In the Word-made-flesh there is a continuity between divine and human, eternal and temporal. Spirit interpenetrates matter. The matter assumed by the Word of God as he became flesh is common ordinary human flesh and blood. On the atomic and sub-atomic level, the matter constituting Christ's body is identical to the matter that constitutes our own flesh and blood.

In the great kenotic event of the incarnation, the Word of God emptied himself to become flesh and dwell among us. This first self-emptying was only the beginning. At the last supper when Jesus blessed bread and said, "Take it, this is my body" (Mk 14:22), and when he handed over the cup of wine saying, "This is my blood" (Mk 14:24), he was emptying himself still further into materiality.

Like the incarnation itself, the eucharistic bread and wine are further signs of God's enduring love for the world: "This is how God loved the world" (Jn 3:16). The Word that first appeared as flesh now appears as bread and wine. The eucharist is the ultimate epiphany, or appearance, of God as purely material reality, in that *world* which God so loves.

Although the whole material world is a sacrament of divine presence and love, and a visible means of attaining communion with God, the eucharist stands out as the sacrament par excellence. The eucharist is that visible point in the cosmos toward which the crucified Christ—now that he is "lifted up from the earth" in glory (Jn 12:32)—is constantly drawing everything from every corner of the universe.

Bread and Wine

The tangible, material elements of bread and wine contain and mediate the real presence of Christ in the eucharist. Here, material reality is transformed and becomes a vehicle of divine life and love. Two common material elements, household items, support a special presence of the sacred.

In designating which food would contain his presence, Jesus did not choose something out of the ordinary. He did not insist on something exotic or imported. Instead he was content to use the bread and wine of the Passover meal—ordinary, everyday items. The bread could be made at home; the wine was readily obtainable if not homemade. These were the staple foods of the populace at that time, their common daily nourishment. Jesus lifted these elements out of their everyday use and gave them a unique significance by transforming their innermost substance into his own substance.

In transforming common bread and wine into his sacred body and blood, Jesus draws our attention to the sacredness of ordinary food and its potential for transmitting and nourishing life. Our daily food contains the germ of life that nourishes our bodily vitality, while the eucharistic bread and wine nourish the life of our spirit. Bread, wine, and all foods are gifts from the giver of life. The germ of life in these holy gifts comes from God and is to be received with gratefulness and reverence.

Communion

As the eucharistic food nourishes our spirit and our body, it unites us with Christ and with all creation. St. Irenaeus, writing in the second century, says in his treatise *Against Heresies:*

> We are Christ's members and we are nourished by creation, which is his gift to us, for it is he who causes the sun to rise and the rain to fall. He declared that the chalice which comes from his creation was his blood, and he makes it the nourishment of our blood. He affirmed that the bread which comes from his creation was his body, and he makes it the nourishment of our body.[1]

By transforming common food into his own body and blood, Jesus tells us that he desires to be our source of life, the life of our life. Bread and wine suggest a meal, and a meal suggests sharing life, celebrating unity and fellowship. Bread is called the staff of life because of its power to sustain the body's strength. Offering bread to someone symbolizes hospitality and friendship. Wine, especially red wine, is a natural symbol of blood and therefore of life. To share a cup of wine with someone is to share life and joy. The alcoholic content of wine increases the enjoyment. The psalmist praises God for "bread from the earth and wine to cheer the heart" (Ps 104:15).

When we consume the eucharistic bread and wine we become one spirit with the Lord Jesus. We are no longer separate, no longer alienated from one another. "Anyone who attaches himself to the Lord is one spirit with him," says St. Paul (1 Cor 6:17). This union comes about through the personal and rather intimate act of taking food and drink into our mouth, swallowing it, and letting it be absorbed until it becomes part of one's own self. However, since Christ is greater than we are, in our eucharistic communion Christ is assimilating us as members of his spiritual body. Without losing our own identity we begin to live by his life, and he becomes the life of our life. As Jesus says,

> "If you do not eat the flesh of the Son of man
> and drink his blood,
> you have no life in you." (Jn 6:53)

The Wheat and the Grapes

The eucharistic bread and wine connect us both with the sacred and with all creation, in an unbroken continuum. The connection is clear if we trace these elements back to their original forms. Bread originates from ripe grain, and wine from the fruitful vine. The growing grain and ripening grapes draw moisture and nutrients from the soil and energy from the sunlight. By their root systems these plants are connected with the earth itself, and by their stems and leaves they are part of the oxygen and carbon dioxide cycle that sustains all living things.

When we trace the wheat and grapes back to their constituent forms on the level of atoms, we reach a zone where all is in flux and where, according to quantum physics, an individual atom or part of an atom may move from the structure of one form to that of another in the intricate dance of possibilities. Everything connects to everything else in this incredibly complex network. In fact, the atoms and molecules that constitute the bread and wine we use today are indistinguishable from those that made up the loaf and the cup that Jesus blessed at the Last Supper.

Wheat and grapes have to be cultivated, harvested, and processed by human hands before they become elements for eucharist. Thus the finished products of bread and wine symbolically gather into themselves not only the earth, the sun, and the elements of the cosmos but all human activity as well.

In this sense all creation participates in the eucharist. All this is the body and blood of Christ. This sanctified creation calls for reverent handling, without waste or abuse—the same kind (though not the same degree) of reverence with which we handle the sacred elements of the eucharist.

A eucharistic prayer from the end of the first century shows that the first Christians made a clear connection between the sacramental bread and the stalks of wheat growing on the hillside. The prayer stresses the unity of the body of Christ in its many members:

> As this broken bread was scattered upon the mountain tops
> and after being harvested was made one, so let Thy church be

gathered together from the ends of the earth into Thy kingdom, for Thine is the glory and the power through Jesus Christ forever.[2]

Sense of the Sacred

If our celebration of the eucharist is fully credible and our symbols are fully convincing, they will have the effect of awakening and nurturing our sense of the sacred as present in the created world. Nature, in her awesome power and beauty, reveals something of the mystery of Christ. We should be able to encounter the risen Christ in nature, as Mary Magdalen encountered him in the garden on the first Easter morning (Jn 20:14).

A sense of the sacred is not nourished by our typical, American, consumerist approach to life or by our compulsion for haste. The eucharist teaches us that the purpose of human life has less to do with accelerated productivity and consumption than with contemplative wonder, love, peace, and joy in the presence of the sacred. The eucharist, properly celebrated, will gradually form in us new attitudes of respect for ourselves, our neighbor, and our common home, which is planet earth.

Toward a New Creation

The creative transformation of bread and wine in the sacrament of the eucharist promises a further transformation of the material world, and even the transformation of human society in justice and peace. The eucharist not only promises, it empowers this transformation. Still, the effect is not automatic but utterly dependent on the work of human hands and human minds, and on the loving generosity of human hearts.

The transformation set in motion first by the incarnation of the Word and then by the sacrament of the eucharist will continue until people everywhere on this planet enjoy a just share of the earth's fruits and the products of human industry and culture. For example, an initial transformation comes about when an immoderate consumer has a change of attitude and begins to use only what is necessary for the present, so that others too may have enough. When God's chosen people lived on manna in the desert, each one collected what was necessary for the day and no one went hungry: "Each had collected as much as he needed to eat" (Ex 16:18).

The transformation will not be fully complete until our planet ceases to be polluted and exploited, until it begins to be healed of the wounds inflicted upon its soil and oceans and atmosphere. The dynamism of the eucharist ultimately culminates in the New Creation. Indeed, Robert Brungs has aptly called the eucharist "the sacramental center of the New Creation."[3]

Conclusion

The eucharist brings about this new creation in the kingdom of God by the transforming power of the risen Christ. Pope John Paul II says in *Sollicitudo Rei Socialis*, an encyclical letter devoted to social and ecological problems:

> The kingdom of God becomes present above all in the celebration of the sacrament of the Eucharist. . . . In that celebration the fruits of the earth and the work of human hands—the bread and the wine—are transformed . . . through the power of the Holy Spirit and the words of the minister, into the body and blood of the Lord Jesus Christ.[4]

The risen Christ acts through those who receive the eucharist and commit themselves to follow his example. His giving of his life for us is the model we are to follow in laying down our life for the healing and freeing of the whole creation (human and nonhuman), currently in slavery to corruption. John Paul II goes on to say that we receive from the eucharist "the strength to commit ourselves ever more generously, following the example of Christ, who in this sacrament, lays down his life for his friends."

In the new creation that is slowly taking shape in the course of history, humanity will no longer approach the environment to exploit it destructively. To do so is sacrilegious, the deliberate ruin of something sacred and beautiful. Reverence for the fruit of the earth and the fruit of the vine will make us sensitive to the interconnectedness of all things and to their potential for transformation. In a eucharistic vision such as this, we have the theological basis for a viable and effective Christian ecology.

Notes

1. Irenaeus, *Against Heresies*, book 5, chap. 2, as found in *A Word in Season III* (Riverdale, Md.: Exordium Books, 1983), p. 97.
2. *The Didache: Teaching of the Twelve Apostles*, trans. by Francis Glimm, in *The Fathers of the Church: The Apostolic Fathers* (New York: CIMA Publishing, 1947), p. 179.

3. Robert Brungs, *You See Lights Breaking Upon Us* (St. Louis: Versa Press, 1989), p. 136.

4. John Paul II, *Sollicitudo Rei Socialis* ("On Social Concerns") (Washington, D.C.: United States Catholic Conference, 1988), art. 48.

12

Ecological Resources
in the Benedictine Rule

TERRENCE G. KARDONG, O.S.B.

A casual reading of the Rule of St. Benedict indicates little or nothing of ecological interest. Written in sixth-century Italy, the Rule comes from a time when the unraveling of civilization seemed to be a far more pressing concern than the preservation of the natural environment. The collapse of the Roman Empire along with the incursion of the northern tribes made St. Benedict's type of localized monastic community valuable as a social unit. Unlike Egypt and Syria, where monasticism had often involved flight from city to desert, cenobitic monasticism in the West would prove to be one of the anchors of the new feudal society.

Ancient monasticism in general was not known for its devotion to the beauties of nature. The Egyptian hermits lived in the desert because they thought it was devoid of beauty and therefore would force them to look inward to the health of their souls. No such negative philosophy of nature can be inferred from the Rule of St. Benedict, but it is also certain that Benedict was no nature mystic. Unlike St. Francis, Benedict is not given to ecstatic expressions of kinship with the natural world. In the Rule, nature is simply taken for granted as the world in which the monks live.

Nevertheless, Benedict has still come in for some praise from ecologists. René Dubos prefers Benedict's advocacy of the stewardship of natural resources over Francis's poetic appreciation of the earth.[1] As a matter of fact, Benedict only uses the term "stewardship" once (64.7), and then in a biblical quote that has more to do with human community than with nature conservancy. We will note a few other texts where stewardship is implied, but they are not many.

One of the weaknesses of the stewardship model is its assumption of human superiority over the rest of nature. This has caused some writers to prefer a philosophy of companionship with nature.[2] Such an attitude is incompatible with the hierarchical mindset of Benedict, but that does not mean that his worldview has no potential for ecological responsibility. Near the heart of Benedict's spirituality lies an acute sense of creatureliness. He seems to be almost overwhelmed by the majesty of God; he is an utterly religious person. From his point of view the world belongs to God and that is the reason why we must love it and care for it. Throughout human history, such an attitude has always been the primary motivation for ecological awareness.[3]

In order to appreciate the position of someone like Benedict in regard to this question, we have to place him in his proper context. One of the gravest heresies of early Christianity was an exaggerated dualism that exalted the spirit at the expense of material creation. Although the height of the crisis had abated by the fourth century, when monasticism came into being, the asceticism of the first monks sometimes spilled over into disdain and even hatred of the body and the world.[4]

There is little or none of this in the Rule of St. Benedict. Of course he is basically concerned with promoting spiritual growth, but he never does it at the expense of the body. Throughout the Rule there is a balanced and realistic attitude toward life that appreciates the importance of the physical. Thus when he legislates for food and clothing, Benedict takes into account the climate and the season (40.8; 55.7). These considerations seem elementary to us, but they are often overlooked by monastic zealots.

Still, it must be admitted that everything that Benedict says about nature is strictly tied to human well-being. While he displays no alienation from nature, he shows no interest in it as a phenomenon in its own right. For some modern ecologists this might classify him automatically as part of the problem and not part of the solution, but the truth is that all ecological thought has to take the human into consideration. The question is: What kind of human life is in question? Is it one that the planet can sustain, or is it one that will ultimately destroy the ecosphere? The contention of the rest of this chapter is that Benedict's monks, if they live according to his teachings, are friends of the planet and not its enemies.

Humility

Humility is probably the central virtue promoted by the Rule. Its importance to the author is suggested by the sheer bulk of chapter

7, where it is discussed in the form of an elaborate ladder image. An examination of the steps of the ladder of humility shows that it also incorporates the virtues of obedience and silence, which are very close to humility in the thought of the ancient monks. The modern mentality tends to distrust all three of these values due to their apparent passivity and threat to human self-esteem.

This reaction is not likely to subside when one examines the theological heart of the chapter on humility, which is exceedingly tough in its outlook:

> The first step of humility, then, is that a man keeps the fear of God always before his eyes and never forgets it. He must constantly remember everything God has commanded, keeping in mind that all who despise God will burn in hell for their sins and all who fear God have everlasting life awaiting them. (7.10-11)[5]

No doubt a modern theologian would put it in different terms, but "fear of the Lord" will always be a daunting idea. The God-fearing person finds the holiness of God so awe-inspiring that it produces a strange mixture of simultaneous attraction and terror.[6] Someone who has experienced the living God knows firsthand that God is God and that he or she, the human subject, is not.

The Christian name for this new awareness is humility,[7] and the etymology of the term reveals its connotation: *humus* means "soil." Humble people freely acknowledge that they are not the Creator of the universe, but merely creatures. In that sense they are one with the soil, the plants, the animals, all of whom "fear God" by their very existence. The human choice, of course, is to live in obedience to their Creator or to reject the constraints of creaturehood and to attempt to play god.

Sometimes Benedict expresses human creatureliness in very harsh terms:

> The sixth step of humility is that a monk is content with the lowest and most menial treatment, and regards himself as a poor and worthless workman in whatever task he is given. . . . The seventh step of humility is that a man not only admits with his tongue but is also convinced in his heart that he is inferior to all and of less value. (7.49,51)

Obviously, such a statement is not consonant with ordinary modern humanistic wisdom. For many people it merely corroborates a lack of self-esteem. One cannot deny the psychological dangers of this kind of language, but at the same time we must find some way out of our current need to hear everything with psychological ears.[8] If we concentrate on the strictly theological content of these verses, they are simply stark statements of basic religious and Christian faith: We come from God and we go back to God; therefore we owe everything to God. Any dilution of this understanding opens the door to human hubris.

Yet it must be admitted that this same spirituality can be used in a sinful way to subject others to an inferior position and to keep them in line. There is no doubt whatsoever that the rich and the powerful have often succeeded in coopting the Christian message and perverting it to their own interests. One of the main purposes of the Enlightenment was to emancipate the human spirit from this kind of religious bondage. Consequently, the physical science and technology that was the fruit of this liberation had a predisposition to reject all appeals to humility and fear of the Lord. In recent times more scientists are coming back to religion through their own discoveries of the wonder of the universe, but the old split between religion and science has not completely healed.

To return to Benedict's doctrine of humility, it should be pointed out that it seems largely aimed at the individual. This is primarily because of the literary history of the text,[9] but it need not stop us from extrapolating to the corporate dimensions the humility. In fact, there is something inherent in human social groups that they resist application to themselves of the very same virtues they press on the individual.

Thus there have been Benedictine monasteries full of humble individuals that nevertheless vaunted their corporate power through every means possible. Nowadays, national governments and multinational corporations seem to hold it as a veritable principle to promote their own advantages at the expense of everything and everyone else. Somehow, what seems proper for the individual appears to be inapplicable for the group. This corporate hubris also seems to coexist quite comfortably with expressions of piety that have little to do with traditional fear of the Lord.

Corporate arrogance poses a serious difficulty for the future of the ecological movement. More and more, individuals seem to be coming to heightened awareness of their own dependence on the ecosystem, and some of them are striving for correspondingly modest lifestyles. Yet the problem is so vast that it will require the kind of changes that can only be effected by governments and corporate industry. Until these entities can transcend the temptation to indulge in self-aggrandizement, the future does not look bright.

Still another feature of Benedict's teaching on humility needs to be taken into account and perhaps remedied. By and large, his kind of humility only seems to demand respect for those higher than oneself. Little is said in chapter 7 about the need to honor and cherish those perceived to be beneath oneself.[10] It should be added at once, though, that the rest of the Rule preaches the need to reverence Christ in the weakest and most marginal persons we meet. In fact, this is an important corollary to Benedict's doctrine of the fear of the Lord: One must fear God who dwells in the little ones of the earth.[11]

While Benedict does not explicitly do so, there is no reason this attitude of compassion and care toward the "lower" cannot and should not be extended to all creatures. A respect for all creatures has always been a central part of the faith of certain Eastern religions, such as Jainism and some forms of Buddhism, but it has received little attention in Christianity. If humility is to be given full expression, however, the human person must not only be humble before God but also before the merest living member of God's creation.

Stability

In the initiation and formal admittance (called "profession") of a novice laid down by St. Benedict in chapter 58, the term "stability" occurs twice (58.9; 58.17). It is one of the three promises or vows the aspirant must make before entrance into a Benedictine community. Although the root meaning of the word *stability* refers to "the ability to stand or stay put in one place," this is not the full extent of the meaning of stability in the monastic tradition.

Among the earliest monastic hermits in the Orient, stability meant the ability to keep to one's hermitage: "Take care of your cell and your cell will take care of you." The purpose of this bit of desert wisdom went beyond preventing aimless wandering by monks, who were prone to it due to their lack of family and possessions. The monk was to keep to his cell in order to focus his spiritual life and also to face those aspects of himself that needed further conversion. Among the cenobites of the West, stability had the added meaning of perseverance in the monastic state until death.

Nevertheless, Benedict puts a good deal of emphasis on the sheer need to stay home, sometimes called *stabilitas loci* or stability of place. Thus he directs that the monastery be a self-sufficient socio-economic unit, containing "water, mill and garden" (66.6-7). The reason given is purely spiritual: "Then there will be no need for the monks to roam outside, because this is not at all good for their souls." What is more, he engages in a rather bitter diatribe against the *gyrovagi* or wandering monks who "never settle down, and are slaves to their own wills and gross appetites" (1.11).

Of course the static ideal could only be kept more or less perfectly, depending on the circumstances. We know that Benedict's monks traveled, since he speaks about their conduct on the journey (50.4) and especially about their conduct upon return (67.1-7). Those returning from a journey are forbidden to talk about what they have seen "because that causes the greatest harm" (67.5). It seems that the legislator is trying to create a closed environment, free from the contamination of alien contacts and customs. This has its limits, of course, and Benedict was wise enough to take advice from visiting monks (61.4) and from reliable local Christians (64.1-6).

When we read these texts, it is hard to imagine a social system less akin to the current post-industrial age in which we live. In our world people not only travel thousands of miles at the drop of a hat, but they cheerfully change their domiciles about every three years on average. Furthermore, world news is beamed constantly into the remotest hamlet, so that everyone knows everything instantly. Even monasteries trying to live the Benedictine Rule today do not and cannot live in isolation from their neighbors.

Nevertheless, the Benedictine monastery was the perfect institution for the age in which it arose. At that time the communication system so carefully created by the Romans was disintegrating, leaving people fragmented in small, isolated, rural communities. In such a situation the quintessentially local, self-sufficient Benedictine monastery was an ideal nucleus of religion, culture, and even commerce. Yet the flowering of the monastery as a social phenomenon was limited. As soon as Europe recovered its former network of roads and cities, the church needed a different form of religious life, and she found it: the friars. It was the hallmark of the Franciscan and Dominican preachers that they traveled from city to city or to wherever the people were to be found.

Our real point here, however, is simply that some degree of physical stability is of vital importance in shaping human attitudes toward the earth. Perhaps this idea could be translated into the following axiom: Those who live in a place have the biggest stake in it. This is in turn a corollary to the ecological truism that degradation of our environment inevitably leads to degradation of ourselves.

But there is danger in generalizing too much in this matter. We are not speaking here of the earth as a whole, although it is obviously a single physical entity in which all parts are interconnected. For the ordinary person, the earth is too big to be understood well on a global level. To really get to know and love a place, a person must live there and live there for a long time. Furthermore, those who live in a place are usually in the best position to know what is appropriate for that place in terms of human initiatives.

It is, of course, easy enough to bring forth examples of how natives have ruined their local environment. Human ignorance can wreak havoc in any situation, and local greed is as bad as any other greed. Furthermore, only science can give us the answers to many large-scale problems. Nonetheless, more mischief has been perpetrated by bureaucrats at long range than by the people "on the land." A glance at the disastrous history of the national planned economy of Soviet agriculture should be proof enough of that. By and large, the best care of the land will come from people who regard it as home. The farmer who is determined to leave the land to his children better than he found it is a far better environmentalist than the conglomerate that regards the land as a purely economic commodity.[12]

To return finally to the Benedictine monasteries. Whatever other adaptations they have had to make to their time and place, they have remained faithful to the localism taught by Benedict. Often this has involved a kind of provincialism that comes with a refusal to pull up stakes and move to the nearest city. Nor can it be said that monasteries throughout history have always been paragons of responsible stewardship. Yet they have enduring value as a witness that a certain stability is necessary to proper care of the earth.

Frugality

In monastic jargon, *frugality* is known as "poverty." This is something every monk vows to live, but it does not mean that one actually lives like the very poorest members of society. Such was the Franciscan ideal, and Francis and a few of his followers kept to it; but both monks and friars have usually been educated, and educated people are never truly poor. Benedict rarely speaks explicitly of poverty, and when he does, he is referring to those who come to the monks for help.

Yet there is a solid ideology of frugality in the Rule, stated succinctly and elegantly in chapter 34: "Whoever needs less should thank God and not be distressed, but whoever needs more should feel humble because of his weakness, not self-important because of the kindness shown him" (34.3-4). The context is Benedict's demand that distribution in his monastery should be made solely on the basis of need.

He is realistic enough to know that different persons have considerably different physical needs. To satisfy those needs may give the appearance of gross inequity, but that is the only real solution to the problem of proper distribution Of course, such an approach could only work in a small community of fully committed persons.

In Benedict's system, this community is lead by the abbot, who must have the discretion and compassion necessary to know what each member needs.

What are Benedict's reasons for such an unusual system? On the one hand, he wants to combat avarice, that universal vice that can never get enough of this world's goods. The avaricious person confuses wants with needs, but Benedict knows that wants are actually insatiable unless they are held in check. Thus he condemns all private property for monks as a "vice to be torn out by the roots" (33.1; 55.18). Of course, he knows that private property is not evil for non-monks, but they too are susceptible to avarice. Has any vice flourished more spectacularly in the late twentieth century?

The second reason for frugality is more communal in nature. If all receive what they need, then there should be "peace among the members" (34.5). This peace will not be merely psychological contentment but something based on the more solid reality of objective satisfaction of legitimate needs. Where that exists, there is a good basis for communal harmony; where it does not, there may be a veneer of calm but violence smolders underneath. "If you want peace, establish justice" (Pope Paul VI).

In Benedict's administrative system, the official in charge of taking care of the physical needs of the members is called the cellarer. Actually, this responsibility falls ultimately on the abbot, but the cellarer is to be "like the father to the whole community" (31.2). As such he must have many of the same characteristics as the abbot, but Benedict also twice insists he be *non prodigus*—not wasteful (31.1,12). And he specifies this in a memorable saying: "He will regard all utensils and goods of the monastery as sacred vessels of the altar" (31.10).[13]

The importance of treating the ordinary things of the monastery with the same reverence as the extraordinary things receives further elaboration in chapter 32, concerning the tools of the community.[14] Here Benedict demands that the abbot himself keep track of the tools and make sure that they are given out and returned in good condition. Anyone who has lived in a commune realizes the potential for abuse when everyone owns everything—and no one considers anything his or her own. For Benedict, tools seem to be symbols of the material world, which needs and deserves our best care and attention.

It must be admitted, though, that the stewardship of things is not a major emphasis for Benedict; the care of people receives much more development in chapter 31 and throughout the Rule. Yet there should be no conflict between these two things. In fact, one gets the impression that the care of people requires close attention to the things that people need and use. When discussing clothing, for example, the Rule says that "the abbot ought to be concerned about the measurements of these garments that they not be too short but fitted to the wearers" (55.8). There are many passages like this in the Rule showing that the little practical necessities of life must be well taken care of if the community is to flourish spiritually.

Perhaps we have strayed a bit from our subject of frugality. Certainly there was a strong ideology of abstemiousness among the monks from the very first. Among the desert monks this ascetic attitude toward material goods sometimes took extreme forms, but it was recognized by thoughtful monastic writers that abstinence could only be sustained if it were not carried too far. Like the modern weight-watchers, Cassian argued for a steady diet that one kept to, whether it seemed too little or too much at the moment.

Benedict's ascetical ideas are generally marked by moderation and good sense more than by towering idealism. Chapter 40, on the amount of drink, is a good example. He notes wryly that monks really should not drink wine at all (40.6), but since they cannot be convinced of this, let them at least do so moderately. On a more serious note, though, is his comment that they should be satisfied if local circumstances make it hard or impossible to obtain wine. What is ecologically important is that one live on what is easily available in the locality.[15]

The same principle is invoked in regard to clothing. Monks are not to "complain about the color or coarseness of all these articles, but use what is available in the vicinity at a reasonable cost" (55.7). This may seem like elementary prudence for those who wish to live a simple life, which ought to be a virtual definition of monasticism. Yet other religious ideals can intrude in such a way that "poor" monks sometimes get caught up in a desire for extravagant church vestments and so on.

The simple, frugal lifestyle extends to many areas of daily existence. One of the most telling signs of poverty in ancient times was the necessity of doing manual work. The upper classes did not sully their hands with this kind of labor, which was thought proper only for slaves. Nevertheless, Benedict tells his monks to do their own harvesting when necessary and not to grumble about it (48.7-9). This, of course, implies that they sometimes hired others to do their farming, but at least they are not to think they are above such things.

In summing up this discussion of Benedictine frugality, we should recall our earlier claim that there is no sign of anti-materialism and there is certainly no hint that matter as such is evil and something to be despised. Of course, the opposite tendency of hedonism is also absent, as one would expect in a monastic Rule. But it was not so easy for a monastic writer to steer clear of excessive asceticism, as is abundantly clear to anyone who surveys the ancient monastic Rules.

In our study of Benedict's attitude toward the physical world, we have not been able to point to any explicit philosophy that might qualify today as adequately ecological. By and large, his great concern is with the spiritual health of a human community. Still, he knows full well that the spiritual can only rest on the material. His remarks about the latter are mostly casual asides, but they are more impressive for all that, since indirect glimpses sometimes tell us a good deal about someone's deep convictions.

Notes

1. See René Dubos, *A God Within* (New York: Scribners, 1972), pp. 168-72.

2. See, for example, Kenneth and Michael Himes, "The Sacrament of Creation: Toward an Environmental Theology," *Commonweal* (January 26, 1990), pp. 42-49.

3. Glenn Tinder mainly emphasizes human compassion for the weak and poor, but the thesis can easily be extended to torn and bleeding nature. See "Can We Be Good without God?" *Atlantic Monthly* (December 1989), pp. 68-85.

4. See Terrence G. Kardong, O.S.B.,"The World in the Rule of Benedict and the Rule of the Master," *Studia Monastica* 26 (Spain, 1984), pp. 185-204. The "world" (*saeculum*) that is meant in these texts is basically human civilization, but misanthropy often goes hand in hand with fear and hatred of the material world.

5. All quotations given here are from *RB 1980* (Collegeville, Minn.: Liturgical Press, 1981).

6. See Rudolf Otto, *The Idea of the Holy* (Oxford, 1923).

7. Pierre Adnés, "Humilité," in *Dictionnaire de Spiritualité* (Paris: Beauchesne, n.d.) 7.1136-87. Adnés shows that humility as such is not found in Greco-Roman thought; it is a distinctive creation of Jewish and Christian theology.

8. Raymond Pedrizetti provides a useful study of the psychological issues involved in the virtue of humility. See *Humility and the Human Ego* (Collegeville, Minn.: St. John's Abbey, 1984).

9. The Rule 7 is heavily based on the *Rule of the Master*, chapter 10 (trans. L. Eberle [Kalamazoo, Mich.: Cistercian, 1977]), which is in turn an expansion of Cassian's *Institute* 4.39 (trans. E. Gibson, in *Nicene and Post Nicene Fathers XI* [Grand Rapids, Mich.: Eerdmans, 1964]). In general, the first seven chapters of the Rule are most heavily marked by the individual, ascetical influence of Cassian.

10. Klaus Wengst shows that the originality of the biblical concept of humility lies in the yoking of two concepts: 1) meekness before authority (*tapeinos*) and 2) compassion for the weak (*praüs*). Especially good is

Wengst's analysis (Chapter 4) of Matthew 11:28-30, "Learn of me [Jesus], for I am meek and humble of heart." See *Humility: Solidarity of the Humiliated* (Philadelphia: Fortress, 1988).

11. See Terrence G. Kardong, O.S.B., "Benedictine Spirituality," *New Dictionary of Catholic Spirituality* (Collegeville, Minn.: Liturgical Press, 1993). I locate the theological center of the Rule in the fear of the Lord (7.10-13). I also show that the fear of the Lord is the basic motivation for compassion throughout the Rule. See also my special study "The Biblical Roots of Benedict's Teaching on Fear of the Lord," *Tjurunga* 43 (Australia, 1992), pp. 25-50.

12. See Wendell Berry, *The Unsettling of America* (New York, 1977). The same basic ideas are found in "Strangers and Guests" (Statement of the Midwest Catholic Bishops on Land Issues [Sioux Falls, SD, 1980]).

13. See André Borias, "The Benedictine Cellarer and His Community," *The American Benedictine Review* 35 (1984), pp. 403-21.

14. See J. Sutera, "Stewardship and Kingdom in RB 31-33," *The American Benedictine Review* 41 (1990), pp. 348-56.

15. Jack Nelson, *Hunger for Justice* (Maryknoll, N.Y.: Orbis Books, 1984).

13

Conversation with the Cosmic Christ

The Spiritual Exercises *from an Ecological Perspective*

WILLIAM J. WOOD, S.J.

A machine using lots of electrical energy . . . that is not grounded poses a serious threat; similarly, a person who is not "grounded" in body as well as mind, in feelings as well as thoughts, can pose a threat to whatever he or she touches. We tend to think of the powerful currents of creative energy circulating through every one of us as benign, but they can be volatile and dangerous if not properly grounded. . . .

We seem increasingly eager to lose ourselves in the forms of culture, society, technology, the media, and the rituals of production and consumption, but the price we pay is the loss of our spiritual lives.[1]

Humankind, having come forth from the womb of the earth just a short while ago in cosmic time, has come to a crucial turning point. Either we will change our ways radically, or we will continue to transform the planet into the hell that it has already become for hundreds of millions of people and the millions of other species who are unable to eke out an existence on this planet, who have no place to call home.

Two millennia have passed since Jesus of Nazareth emerged from the wilderness to announce the good news that, if we undergo a radical change in our attitude, we will be able to make our way through the narrow gate and over the hard road that leads to the home the Creator has been building for the whole creation since the beginning of time.

In the course of those twenty centuries, many ways of following Christ have arisen and passed the test of time. Among them, in the

Catholic heritage, are the traditions associated with names such as Benedict, Dominic, Teresa of Avila, Margaret Mary Alacoque, Hildegaard of Bingen, Francis, Clare, Vincent de Paul, and Catherine Macauley.

The ancient spiritual ways have perdured through the ages, not because they embody a classical moment of arrival at absolute truth, but like the good news itself, they were born of the experience of the living Spirit that constantly makes things new. Like the whole of creation, whose health manifests itself in the limitless diversity of its unfolding, the wisdom of the ancients reveals its mettle when it embraces the freshness of today to generate hope for tomorrow.

This chapter presents the spirituality introduced in the sixteenth century by the Spanish Basque noble Ignatius of Loyola. Like every authentic spirituality, the Ignatian "way of proceeding" issued from Ignatius's direct experiential knowledge of utter Spirit, which transformed the way he lived out his life in the world.[2] It was as a lay person that Ignatius reached the peak of his mystical experiences. And it was for lay people that he compiled the *Spiritual Exercises*, though Ignatian spirituality has shaped the lives of countless professed religious and clerics, as well as lay people, over the past four and a half centuries.[3]

Ignatius Loyola lived in an age of crisis like our own, rocking with conflict and uncertainty, yet full of promise.[4] One world order was crumbling and a new one was struggling to be born. Voyagers dreamed of new worlds and of journeys to the end of the earth, while bloody wars tore Europe apart. Corruption was rife in the church, as well as in civil society, the two pillars of human culture at that time. Christendom, which had prevailed since the third century, was suddenly coming apart at the seams. The world staggered in culture shock, disoriented and disillusioned. Insecurity led to selfishness and further disregard for the common good.

In the midst of this confusion, people yearned for understanding, meaning, and wholeness. The sensational technological and cultural advances overwhelmed them. At the same time, they could not help but notice that the consequences of making progress without reference to the will of the Creator were devastating. The problem, in the sixteenth century as today, was not progress but the human arrogance that denies accountability to the rest of creation and to the Creator, thereby transforming progress into disaster and frustration.

In that scene we find a clever, womanizing, and ambitious knight, born "Iñigo de Loyola," serving at the Court of Castille. Bold in the service of his king, he led a desperate attack, which almost killed him, in the battle against the French at Pamplona. One leg almost totally shattered, the other seriously injured, he was carried home over the mountains to the Loyola castle. His vanity drove him to have his leg broken again and reset, so he would not look unattractive at court when he appeared in doublet and hose.

But during his long convalescence, something happened that turned the life of Iñigo upside-down. His own account reveals the emergence of what is perhaps most original in Ignatian spirituality, the intuition that, *if only we seek, we will find God present at every moment in every thing.* Ignatius, as he called himself after his conversion, tells the story in his autobiography, speaking of himself in the third person:

> As he was much given to reading worldly books of fiction, commonly labeled chivalry, when he felt better he asked to be given some of them to pass the time. But in that house none of those he usually read could be found, so they gave him a life of Christ and a book of the lives of the saints in Castilian.
>
> But, interrupting his reading, he sometimes stopped to think about the things he had read and at other times about the things of the world that he used to think of before.
>
> Yet there was this difference. When he was thinking of those things of the world, he took much delight in them, but afterwards, when he was tired and put them aside, he found himself dry and dissatisfied. But when he thought of going to Jerusalem barefoot, and of eating nothing else but plain vegetables and of practicing all the other rigors that he saw in the saints, not only was he consoled when he had these thoughts, but even after putting them aside he remained satisfied and joyful.
>
> He did not notice this, however; nor did he stop to ponder the distinction until the time when his eyes were opened a little, and he began to marvel at the difference and to reflect upon it, realizing from experience that some thoughts left him sad and others joyful. Little by little he came to recognize the difference between the spirits that were stirring, one from the devil, the other from God. From this lesson he derived not a little light.[5]

What Ignatius stumbled upon was not some esoteric or arcane spiritual discipline open only to the spiritually gifted and theologically trained. The Ignatian way of life pays critical attention to what goes on inside us as we engage in the ordinary activities of life; it lets that contact with God move and shape our action. We, like Ignatius, spend time in prayer, not because that is where we find God, but to reflect and follow up on God's communication with us in our daily activities. Ignatius discovered that every human experience has a religious dimension because the world is filled with the Spirit of God Incarnate in the Cosmic Christ, whose death and resurrection has liberated us from the power of evil and the shackles of sin.

Through his reading and reflection at Loyola, Ignatius determined to make a complete break with his past and inaugurate a new life of imitation of the saints. His spirituality was generous and determined but inexperienced, uninformed, and imprudent. But his thought was already oriented, more than he realized, to cooperation with Christ actively at work redeeming the world. The deepening came in the next stage of his growth.

Though not wholly recovered, he set out after six months of recuperation at Loyola, on a pilgrimage to the shrine of the Black Virgin of Montserrat. After a few weeks of prayer and study in a Carthusian monastery at Montserrat, he spent the next year at nearby Manresa, where his *Spiritual Exercises* began to take shape.

Taking copious notes, Ignatius made an attempt to describe the lights and insights granted him during the intense period of retreat at Manresa. But his descriptions fall short in conveying just what it was he "saw." That "he saw the Most Holy Trinity in the form of three musical keys,"[6] for example, does not tell the reader much, even though it moved Ignatius to tears for days and, as he recorded in his autobiography, "the effect has remained with him throughout his life."[7]

Ignatian spirituality, as we shall soon see, gives us something more valuable than descriptions of Ignatian mystical experiences. It teaches us how to find the living God in our own experience, how to distinguish between God's voice and false voices, and how to live in the Spirit of the Cosmic Christ pulling everything together according to the loving plan of the Creator.

Effects and actions meant more to Ignatius than feeble attempts to describe his spiritual insights in words. This is illustrated by his account of what he regarded as the most definitive of his mystical experiences, the one that occurred while he was out for a walk from Manresa and stopped to rest by the River Cardoner. "While he was seated there," he wrote in his autobiography,

> the eyes of his understanding began to be opened; not that he saw any vision, but he understood and learnt many things, both spiritual matters and matters of faith and of scholarship, and this with so great an enlightenment that everything seemed new to him.
>
> *This left his understanding so very enlightened that he felt as if he were another man with another mind.*[8]

What counted for Ignatius was not knowing much, but being loved to life by Whom he came to know. And so he wanted to map out the trail that led to his life-transforming contact with the living God. Not content to tell others what he saw, he wanted them to experience it for themselves.

So, he compiled a detailed "how to" manual and distributed it almost immediately for the use of others, explaining at the outset precisely what he meant by "spiritual exercises."

> By the term Spiritual Exercises we mean every method of examination of conscience, meditation, contemplation, vocal or mental prayer, and other spiritual activities, such as will be mentioned later. For, just as taking a walk, traveling on foot, and running are physical exercises, so is the name of spiritual exercises given to any means of preparing and disposing our soul to rid itself of all its disordered affections and then, after their removal, of seeking and finding God's will in the ordering of our life for the salvation of our soul.[9]

He provided a succinct table of contents and adjustable timeline for giving the *Spiritual Exercises*. For those who could devote full time to the enterprise, it took about a month. So, Ignatius divided the *Exercises* into four parts, calling each a "Week." The First Week was devoted to the consideration of sin; the Second Week to the incarnation and the life of Jesus up to Palm Sunday; the Third Week to the events of the passion; and the Fourth Week to the resurrection and ascension. For those unable to give full time to the experience, the *Exercises* took several months to complete.[10]

The Ecological Dimension of the *Exercises*

For Ignatius, dialogue between the exercitant (the person doing the *Spiritual Exercises*[11]) and an experienced director[12] was crucial to the effectiveness of the exercises. The need for guidance implies a sense of connectedness and interdependence. Ignatius insisted that we can never be the best judge in our own case, and that we need to seek direction humbly and follow it with critical obedience. In that spirit he inserted a brief "Presupposition" before launching into the First Week:

> That both the giver and the receiver of the Spiritual Exercises may be of greater help and benefit to each other, it should be presupposed that every good Christian ought to be more eager to put a good interpretation on a neighbor's statement

than to condemn it. Further, if one cannot interpret it favorably, one should ask how the other means it. If that meaning is wrong, one should correct the person with love; and if this is not enough, one should search out every appropriate means through which, by understanding the statement in a good way, it may be saved.[13]

The "Presupposition" arose mainly from his experience of being hauled before the Inquisition on a number of occasions, charged with heresy by some disgruntled ecclesiastic who felt that the *Exercises* must be heretical because Ignatius, untrained in theology, had no credentials.

But there seems to be something else at work in the "Presupposition," as well. Throughout the *Exercises* Ignatius manifests a kind of latent sense of what we know today as the basic principles of ecology: 1) the interrelatedness of all beings, 2) the conservation of all resources by living systems, and 3) greater diversity and variation for healthier ecosystems. Can we not already see traces of this threefold ecological premise in the "Presupposition"?

1) Mutual trust must be presupposed because of the whole concatenation of connections from which the exercises spring, to say nothing of the inescapable interdependence between the one giving and the one receiving them. The exercitant needs the guidance of the director to make sure he or she is on the right path, while the director must be aware of and defer to the action of the Spirit, Who moves in its own way, in the life and prayer of the exercitant.

2) At the same time, individuals can be wrong. So, along with tolerance and mutual trust, there is need for vigilance to conserve the truth, to keep us in contact with reality, which includes what comes to us from tradition and the church, as well as from God's action within individuals.

3) But in a healthy living system, conservation will go hand-in-hand with diversification to form the dynamically connected Whole. In doubtful cases, the presumption will always favor the difference. The traces of the divine are limitless and always fresh and new. We can find God, therefore, in all things, including approaches and points of view that differ from our own, and that we may not understand.

To our age, characterized by unspeakable human suffering and devastation of the environment to the extent of putting the planet in peril, Ignatian spirituality offers a double-edged wedge of liberation. The clear-sighted spiritual discipline that the exercitant develops in the course of making the *Spiritual Exercises* constitutes the first edge, the impassioned and critically discerning way of life of one committed to living in the Spirit to play a part in the redemptive mission of Christ in the world.

This edge addresses the Ignatian way as liberation for justice. It prompted theologian Monika K. Hellwig to highlight five features of Ignatian spirituality which make it as suitable for a generation struggling for justice in the space age as for the generation that faced similar challenges in the age of discovery and renaissance:

1. Grounding of everything in profound gratitude and reverence.
2. Continuous cultivation of critical awareness.
3. Confident expectation of empowerment to accept and exercise responsibility.
4. Unequivocal commitment to action.
5. Recognition that the gospel of Jesus Christ is essentially countercultural and revolutionary in a nonviolent way.[14]

My task here, however, focuses on how a motif of eco-justice, quite overt in the First Week and Fourth Week, flows through the exercises to form their other cutting edge.

Three major themes running through the *Exercises*—the order of the universe, the incarnation of God, and unleashing the Love of the Trinity—derive from spiritual experiences Ignatius had at Manresa.

As I indicated earlier, Ignatius could not find words to describe in any meaningful detail what he "saw" during these episodes. But the words he used, particularly to describe the creation of the universe, sound something like descriptions of the Big Bang, traces of which contemporary astronomers are able to see through powerful telescopes. Consider these accounts from his autobiography:

> Once, the manner in which God created the world was presented to his understanding with great spiritual joy. He seemed to see something white, from which some rays were coming, and God made light from this.
>
> One day while he was hearing Mass [at Montserrat] at the elevation of the Body of the Lord, he saw with interior eyes something like white rays coming from above. Although he cannot explain this very well after so long a time, nevertheless, what he saw clearly with his understanding was how Jesus Christ our Lord was there in that Most Holy Sacrament.
>
> Often and for a long time, while at prayer, he saw with interior eyes the humanity of Christ. The form that appeared to him was like a white body, neither very large, nor very small, but he did not see any distinction of members. . . . He has also seen Our Lady in similar form, without distinguishing parts.[15]

World Order and Disorder

Whatever the significance of these experiences, however, it is clear that the major themes of the *Spiritual Exercises* have cosmological and ecological overtones. Having done the exercises over more than forty years, I have grown to reflect on the fact that I do them on three different levels. First, I find myself caught up in conversation with God in a totally personal and individual way, almost as if God and all of creation were there for me alone, as if my salvation and eternal happiness was the most important thing in the world. On a second level, I experience myself entering into conversation with God as the people of California, the citizens of the United States, the global human community—as those whose lives are filled with unimaginable suffering, and those who have access to the power to relieve that suffering and usher in an era of unimaginable peace and happiness. Only in the past few years have I done the exercises from a third perspective, as the voice of Creation crying out in agony to God, like a woman painfully groaning to give birth to a whole new creation (Rom 8:15).

The findings of the physical and social sciences, as well as biblical revelation, support the validity of these three dimensions of the self. The God of revelation is indeed at work in the life of each individual, even every bird of the air and wild flower. Each of us is, at the same time, a hologram of society and of the human species. Like a single small part of a holographic plate, I reflect, no matter how faintly, a representation of the sum total of the values, choices, and assumptions that make up humankind, its paradigms, institutions, and structures.[16]

Moreover, each of us is also a microcosm of the universe, containing all of the universe within us. We are "*Adam* born of '*adamah*,'" as Walter Brueggemann explained at a Theology of Land conference.[17] Connected through earth to the Whole of creation, we, the human, are "a mode of being of the universe," to quote Thomas Berry, "that being in whom the universe comes to itself in a special mode of conscious reflection."[18]

We have our place in the universe, as individuals, as human society, as parts integral to the functioning of the Whole. The exercises put us into conversation with God about all this, beginning with the opening consideration that Ignatius dubbed the "Principle and Foundation," or simply the "Foundation." Laying the groundwork for what is to come, while providing a compact summary of the chief principles of Ignatian spirituality, the "Foundation" presents God's plan in creating humankind and orients us toward fitting ourselves into our role in the order of the universe as it evolves.[19] "Principle and Foundation" ends with this sentence, "I ought to desire and elect only the thing which is more conducive to the end for which I am created." In other words, I contribute to the order of the universe by making sure my own house is in order. When all houses are in order, the universe is in order.

The bulk of the First Week focuses on the disorder in the universe as we know it, caused by sin. Sin is the refusal to accept God's creative plan and hope for us. It is the refusal to live according to our nature as interdependent and loving human beings, whose genuine self-interest consists in working for the common good. It is the refusal to act as creatures, denying our identity and accountability as earthlings.

Meditation on the effects of sin aims to arouse compassion in the exercitant, compassion for the creature harmed by sin, compassion for the Creator, whose creative will has been thwarted by sin. From compassion the exercises of the First Week lead to anger at the injustice wrought by sin; deeper sorrow for our own sinfulness; and gratitude springing from hope in God's will to forgive, redeem, reconcile, and renew the face of the earth.

Understanding the meaning of sin in terms of disorder, which we are coming to recognize as cosmic disorder, gives rise to these passions. Sin offends God within creation, not in the sense that there is any injury to the inner life of the Trinity. Sin defies the divine plan and blocks its realization in the unfolding of the universe, or that which Thomas Berry calls *cosmogenesis*.

The First Week contains three elements that illustrate the cosmic dimension of sin: 1) the analysis that sin causes disorder in the universe, 2) the history of the effects of sin on creation, and 3) the meditation on hell.

Disorder: The harm done by human sin is not so much that it introduces disharmony within the individual person, but that it strikes directly at the order of the universe established by God in Christ. Sin is essentially a refusal to accept the responsibilities of consciousness and freedom, to play our part in overcoming the effects of sin and completing the cosmogenesis willed by God.

The History of Sin: The meditations on the sin of the angels and of Adam have often been proposed as mere preliminaries to our own sin, as if the only thing that really counts is the individual's personal relationship with God. But it seems more accurate to say that Ignatius deliberately set out to show us that personal sins involve not only God and the individual, but the entire universe. Thus, we are connected even in our sinfulness, caught up in the history of sin and its impact on creation.

Although Ignatius, aware that the only one we can change is ourself, tailored these exercises for the use of individuals, internal evidence suggests that, at some level of consciousness, he recognized the individual as a hologram and a microcosm, whatever words he may have found for those metaphors. This is nowhere more obvious than in the Fifth Point of the Second Exercise of the First Week, the meditation on our own sins. The Fifth Point, he explains, consists in:

> an exclamation of wonder and surging emotion, uttered as I reflect on all creatures and wonder how they have allowed me to live and have preserved me in life. The angels: How is it that, although they are the swords of God's justice, they have borne with me, protected me, and prayed for me? The saints: How is it that they have interceded and prayed for me? Likewise, the heavens, the sun, the moon, the stars, and the elements; the fruits, birds, fishes, and animals. And the earth: How is it that it has not opened up and swallowed me, creating new hells for me to suffer in forever?[20]

This is rhetoric, of course, but it is not meaningless rhetoric. Ignatius is driving home the point that our sins are an affront, a personal injury, not to God in naked isolation on the clouds of heaven, but to God in the very mountains.

Hell—The Meaning of Sin: The position of the meditation on hell in the *Exercises* is significant. It is not an independent exercise, not an attempt to terrify us, not even primarily to deter us from sin. By having us use the five senses in our imagination of hell, Ignatius attempts to evoke an understanding of the malice of sin by having us look at the logical consequences of sin. Hell is where sin's own inner orientation reaches its final realization. If sin is the withdrawal from the providence of God, hell is God's acceptance of this withdrawal. Hell is succeeding in sin. The connection between sin and hell is between inner orientation and final realization.

The personal conversations with God or Jesus or Mary that Ignatius calls colloquies provide the best key for what Ignatius had in mind for each meditation in the First Week. It is at this juncture, when he felt that we would be the most well disposed, that Ignatius expresses the full purpose of the exercise. Take the colloquy for the First Exercise, for example. Ignatius doesn't expect us, having immersed ourselves in the history of sin and our own part in that history, to say to Jesus, "Wow, I'm sorry. I'll stop sinning." No, the response he hopes to elicit is a fervent desire to do something to undo the damage done by sin. This is the activist dimension of Ignatian contemplation: *The test of the prayer will be the action in which it results.*

We see the effects of sin on our T.V. screens, if not in our neighborhoods, as Ignatius never saw it. Yet his formula for undoing the ravages of sin, for introducing a new world order characterized by justice and peace and loving relationships, is still applicable today. The *Spiritual Exercises* brings us into contact with the living God, who will show us the way to the new world order Jesus referred to as "the realm of God," the place God calls home, *oikos*, in Greek, the root prefix of *ecology*.

Conversing with the Cosmic Christ

So, Ignatius brings us before the crucifix to confront us with the concrete reality of the world as it is, in contrast with the reality of the world as God created it and intends it to be. Now, we ask—not rhetorically or with grandiosity, but simply, humbly, and sincerely— What have I done for Christ? What am I doing for Christ? What will I do for Christ? In other words, we pray intimately to Jesus Christ about how we can do something about the hellish effects of sin in the world. The Jesus of Nazareth whom we imagine hanging on the cross is the same risen Savior whose saving acts we recapture in the eucharist, the Cosmic Christ, reconciling the world to himself. We want to be part of that! It gets more explicit in the colloquy to the Third Exercise, where we ask for 1) an interior knowledge of sin; 2) an understanding of the disorder of sin; and 3) a knowledge of the world. Finally, the colloquy to the Fifth Exercise, the meditation on hell, once more involves us in a historical context. We speak of hell not only in terms of ourself, but of all human beings, reminding us that our own moral life, our own goodness and badness, is caught up with the history of the entire world.

God's incarnate action in the world captures the focus of the Second Week. Meditation on the story of Jesus of Nazareth provides the shape of the good news of salvation, lending imagination to our direct encounter with the living God in the Cosmic Christ.

The Second Week might seem to present the biggest obstacle to doing the *Spiritual Exercises* from a creation spirituality perspective. Here at the very core of Ignatian spirituality, we are presented with the mysteries associated with the redemption. Considerations of creation could well have been handled in fifteen minutes, according to the interpretation of Fr. Joseph Wall, S.J.,[21] while the need for and the path of redemption occupy the rest of the *Spiritual Exercises*. Jesus Christ, the Redeemer, who died for our sins, is their undeniable focus.

For some deep ecologists, it is precisely such preoccupation with the Redeemer that has led faithful Christians to preside over the devastation of creation. After all, what difference does it make what happens to this world? Getting to heaven is what counts. Though there may be plenty of Christians who in fact think this way, it is a sad misreading of the good news.

The Second Week of the *Exercises* tells a quite different story. There, we get to know a humble, poor, compassionate Christ, who cannot but suffer with the suffering, the incarnate presence and action of God in this world, an earthling of the human kind, with all the genetic coding traced back, not only to David, Abraham, Isaac, Jacob and Adam, but even to the fallout of the first explosion of the Word in the beginning, nineteen billion years ago.

The Second, Third and to some extent Fourth Weeks of the *Exercises* do not present the definitive event of human history as some sort of heartening, let alone sentimental, set of remembrances. Ignatius would have us "enter into" these stories as revealing what Albert Nolan calls the "shape" of the good news of God's here-and-now action in the world.[22] Therefore, along with the contemplations of the biblical episodes themselves—which are the heart of the *Exercises*—there are structural meditations, such as The Call of the King and The Two Standards, as well as intermittent sets of directives, most notably the Rules for Discernment of Spirits.

The strategy of Christ is revealed in the Second Week and contrasted with that of Satan, who lures us to covet riches and start on the destructive path toward honor and pride. Ignatius invites us to consider the address that Jesus, now presented to the imagination as a sort of legendary King Arthur, delivers to those who would follow him:

> He recommends that they endeavor to aid all persons, by attracting them, first, to the highest degree of spiritual poverty and also, if his Divine Majesty would be served and pleased to choose them for it, to no less a degree of actual poverty; second, by attracting them to a desire of reproaches and contempt, since from these result humility.
>
> In this way there will be three steps: the first, poverty in opposition to riches; the second, reproaches or contempt in opposition to honor from the world; and the third, humility in opposition to pride.[23]

Ignatius's appreciation of poverty, revealed in the *Constitutions of the Society of Jesus*[24] as much as in the *Exercises*, rivalled St. Francis of Assisi's love of Lady Poverty. Not only was the prevalent avarice of so many of his contemporary ecclesiastics damaging the church, but Ignatius had already in his first conversion embraced the mendicant spirit of St. Francis, which he had read about during his convalescence. For both saints, moreover, love of poverty went hand-in-hand with a sense of reverence for all of God's creation. If the ecological spirituality of Francis centered on nature, that of Ignatius centered on the universe. Both, however, knew that the living God is always naked, hungry, homeless, poor, and rejected, and that the way to live with God in the world is the way of poverty.

More explicitly than any of the others, the Fourth Week is solidly planted in the place where we live and move and have our being in the real world. Contemplation of the resurrection starts with the historical account but moves rapidly into finding the risen Christ in all things, now. This is the spring from which the Jesuits Gerard Manley Hopkins and Pierre Teilhard de Chardin drank, moving the poet to sing, "Glory be to God for dappled things," and the scientist to observe the Cosmic Christ as the ultimate horizon of the Divine Milieu.

Those who do the exercises come into psychic and spiritual contact with the living God, the Great Spirit who dwells within the rock people and all the community of earth, day by day unfolding the creation. God is alive and active in what we call the material universe, to which our bodies are connected genetically and in the air we breathe. So, we apply the five senses in our contemplations of the Jesus story; God's revelation can be seen, heard, smelled, tasted and touched here and now.

Why did Ignatius not talk in those terms? For one thing, he did not have the scientific information we now have, such as knowledge of genetic codes and the fact of the unconscious. But I suggest that the main reason was that he and his contemporaries took for granted a sense of reality, of connectedness, of nature that we have lost over the last few centuries. The Aristotelian realism of Thomas Aquinas was in vogue when Ignatius conceived of the *Spiritual Exercises*. It was almost a century after Ignatius's vision by the River Cardoner that René Descartes had a vision on the banks of the Danube that would launch the scientific revolution.

In contrast with Descartes's mechanistic world filled with inanimate matter, the world of Ignatius is "charged with the grandeur of God,"[25] and "humankind is one vibrant strand in an elaborate web of life, matter, and meaning,"[26] the web from which the immigrant Word took flesh.

The Third Week brings us to the foot of the cross, in loving communication with the God who, stripped of all divine prerogative, stands faithfully on the side of the poor and oppressed, suffering persecution for the sake of justice.

In this act of dying on the cross, and in the acts of the Last Supper through which the power of the cross was made perpetually available, the God who abided in Jesus of Nazareth rose from the dead to continue living incarnately in the world as the Cosmic Christ. It is this Christ with whom we are conversing in the course of the *Spiritual Exercises* and in the daily way of life that springs from them.

Learning How to Live Like God

Just how we do that, according to the dynamics of the *Spiritual Exercises*, reminds me of the story about a tourist on his first trip to New York City. Searching for Carnegie Hall, he knew he was in the right neighborhood, but not sure of exactly which direction to walk. Just as he was about to pull out the map again, he spotted an old man with a violin under his arm.

Surely he knows the way, thought the tourist. "Excuse me, sir," he called out to the old man, "Can you tell me how to get to Carnegie Hall?"

The old man paused, pursed his lips and replied, "Practice, son, practice."

Moving immediately from "Principle and Foundation" to nitty-gritty practice, Ignatius challenges us to make a rigorous inventory of our lives and of the choices we have made, as well as the choices that lie before us. Like the Twelve Steps of Alcoholics Anonymous, the genius of the *Spiritual Exercises* lies more in their prescriptions for action than in brilliant ideas. The *Exercises* are the "practice" of those who would live in the Spirit. And the first thing required for living in the Spirit is to clean up our act!

Ignatius gives detailed instructions for two types of examination of conscience. The first, comprised of three brief exercises each day, aims to root out particular sins or faults. When we get up in the morning we make a conscious resolve to guard carefully against the particular sin or fault we want to correct or amend. At midday, in a prayerful context, we thoroughly review our performance since rising and make a written record of each time we fell into the particular sin or fault. The second exercise is repeated shortly before going to bed, reviewing the time from the previous examination to the present one.

Not very romantic, is it? Nor is it heroic, earth-shaking, or sensational. But it works. I was reminded of its effectiveness when I went through a Smoke-Enders program and managed to kick the smoking habit on July 1, 1982, after thirty years of addiction and countless failed attempts to quit. A major part of the program was daily conscious examination of what we were doing, thinking, and feeling with regard to smoking, and keeping a written record of each cigarette we smoked each day. It became clear that, among other things, smoking was being pulled out of the mire of habit to the clean air of conscious decision-making, from the level of compulsive addiction to that of free choice.

Already looking to the time after the doing of the exercises, from the outset Ignatius places more importance on methodology than content. Even when he finally gets to the meditation that he entitles the First Exercise (a meditation on three first sins), the bulk of what he has to say has to do with technique ("using the three powers of the soul . . . after a preparatory prayer and two preludes [the exercitant meditates on] three main points [and concludes with] a colloquy," etc., etc.).

He gives detailed advice on using the imagination, the memory, and the five senses. He counsels on how to bring abstract ideas down to earth by letting our imagination make them concrete. For example, to meditate better on the idea of being caught up in sin, it may help to imagine ourself exiled in some hostile land among brute animals. We can be sure that technique worked for Ignatius, and that's why he gets so specific.

Unlike the Twelve Steps, the *Exercises* are fraught with specific details and directions about *how* to work them. There are even directions for scheduling the times and focus of the various meditations. The First Exercise, which would typically be done at midnight, and the Second, which would be done at dawn, introduce new matter. The Third and Fourth are repetitions of the first two, while the Fifth involves applying each of the five senses to the same matter.

Ignatius instructs us to pray at various times of the day and in different bodily positions, whether kneeling, or lying prostrate, sitting, or standing quietly—that is, putting our physical self, posture, state of mind (perhaps most open at midnight or 6:30 a.m.) to the task of praying as fruitfully as possible.

Ignatius was far ahead of his time in attending to certain psychological factors, though he wrote without benefit of the terminology, let alone the clinical experience and scientific studies that have helped shape the culture of this century. Without having a word for it, he knew, for example, that there was a lot going on— and a lot that could be made to go on—in what we now call the unconscious. He knew that what a person read and/or thought about before going to bed at night went through some kind of mental processing during sleep. He recognized the importance of the imagination in bringing ideas to life and the importance of the heart, of passion, in moving people to action.

And Ignatius appreciated the importance of diversity in methodology as well as content, in activity as well as prayer. Even at the very end, as if the exercises had not been replete with different ways of praying already, Ignatius wrote pages to explain "Three Methods of Prayer."

Finally, in the Fourth Week, we learn, not only how to find God in all things, but how to live in the Spirit, which is to live in union with the Cosmic Christ each day, and make the healing action of God available for the healing and restored process of the unfolding of the Whole of creation.

To live in the Spirit in conversation with the Cosmic Christ is to live the life of the Trinity, which is to unleash the power of mutual love—community-building, homemaking, ever-expanding love. So, the main part of the Fourth Week, and the culmination of the *Spiritual Exercises*, which is really their commencement, is an exercise geared to learning how to love as the Trinity loves.

The *Spiritual Exercises* awaken the senses to perceive, open the mind to know, and transform the heart to love the world, which is charged with the incarnate compassion of the living God, who calls each and every one of us to play our part in the unfolding of creation. The response aroused by the *Spiritual Exercises* entails a passionate commitment that Teilhard captured in a prayer he composed, which fittingly concludes this essay:

> O God, I wish from now on
> to be the first to become conscious
> of all that the world loves, pursues, and
>   suffers;
>
> I want to be the first to seek,
> to sympathize and to suffer;
> the first to unfold and sacrifice myself,
>
> to become more widely human
> and more nobly of the earth
> than any of the world's servants.

Notes

1. Al Gore, *Earth in the Balance: Ecology and the Human Spirit* (Boston: Houghton Mifflin, 1992), pp. 220-21.

2. The expression Ignatius most frequently used to describe his spirituality—and that of the Society of Jesus—was "our way of proceeding." See Fritjof Capra and David Steindl-Rast, with Thomas Matus, *Belonging to the Universe: Explorations on the Frontiers of Science and Spirituality* (San Francisco: HarperSanFrancisco, 1991).

3. It may help to dispel the mystique of the *Spiritual Exercises* to realize that most of those whom Ignatius guided were young lay people, ranging in age from eighteen to twenty-five. Not long after his time at Manresa,

because his lack of theological training got him into trouble with the Inquisition, Ignatius decided to get a formal education and eventually went to the University of Paris. In the course of his eight years there he made friends with a number of younger students, each of whom he directed individually through the Exercises. Among the students were Francis Xavier, Pierre Favre, Nicholas Bobadilla, Alfonso Salmeron, and Diego Laynez. In 1540 they would become the first Jesuits, committing themselves with Ignatius to form a new religious order, officially called the Society of Jesus.

4. I am indebted in this biographical section to Father Peter-Hans Kolvenbach, superior general of the Society of Jesus, for his wonderfully synopsized account of Ignatius's "conversion" experience in the keynote address he delivered at the 4th Congress of the World Union of Jesuit Alumni, in July 1991, commemorating the 500th birthday of Ignatius, and the 450th anniversary of the founding of the Society of Jesus. See the World Union of Alumni/Old Boys newsletter, ETC, no. 3 (1991), pp. 7, ff.

5. *Autobiography*, 6, 8, 9, in *Ignatius of Loyola: Spiritual Exercises and Selected Works*, ed. George E. Ganss, S.J. (Mahwah, N.J.: Paulist Press, 1991), pp. 70, 71.

This volume presents the four major writings of Ignatius: the *Autobiography* and *Spiritual Exercises* in their entirety, and his *Spiritual Diary* and *Constitutions of the Society of Jesus* in selections so chosen as to give an overview of each work. It also offers ten samples of his almost seven thousand letters. Ample explanations are given in the introductions and commentaries by way of notes, and there are eight pages of bibliography.

In subsequent notes I will simply indicate the document referred to, with its section numbers and the pages on which they appear in this Ignatian collection.

6. *Autobiography*, 28, p. 79.

7. *Autobiography*, p. 80.

8. *Autobiography*, 30, pp. 80, 81 (emphasis added).

9. *Exercises*, 1, p. 121.

10. *Exercises*, 4, 18, 19, pp. 122, 126, 127. Father Joseph Tetlow has provided a valuable resource for giving and receiving the *Exercises* over a period of months. See Joseph Tetlow, *Choosing Christ in the World: Directing the Spiritual Exercises of St. Ignatius Loyola According to Annotations Eighteen and Nineteen, A Handbook* (St. Louis: The Institute of Jesuit Sources, 1989).

11. For lack of a better word, I will call the one who is going through the experience of the *Spiritual Exercises* the exercitant. Furthermore, to highlight their orientation to action, I will usually describe the action of the exercitant as *doing* the *Exercises*.

12. Ignatius expected that anyone who had done the full *Exercises* was well-prepared to direct others through them.

13. *Exercises*, [22], p. 129.

14. Monika K. Hellwig, "Finding God in All Things: A Spirituality for Today," *Sojourners* (December 1991), p. 12.

15. *Autobiography*, 29, p. 80.

16. See Gore, p. 11. Gore explains the metaphor of the hologram.

17. See Walter Brueggemann, "Land: Fertility and Justice" in *Theology of the Land*, ed. Bernard F. Evans and Gregory D. Cusack (Collegeville, Minn.: The Liturgical Press, 1987), pp. 41ff.

18. Thomas Berry, *The Dream of the Earth* (San Francisco: Sierra Club Books, 1990), p. 16.

19. Notes for the *Exercises*, 18, p. 393.

20. *Exercises*, [60], pp. 139-40.

21. Father Wall never published the contents of his course on the *Spiritual Exercises*, but student notes on his lectures were mimeographed in 1960.

22. Albert Nolan, *God in South Africa: The Challenge of the Gospel* (Grand Rapids, Mich.: Eerdmans, 1988).

23. *Exercises*, [146], pp. 155, 156

24. *Constitutions*, chap. 2, [553], p. 305.

25. Gerard Manley Hopkins, *God's Grandeur*.

26. Vice President Al Gore, although not referring to Ignatius, used this expression in his book *Earth in the Balance*, p. 253.

14

Gaia—Samsara—Narnia

TESSA BIELECKI, O.C.D.

There are many schemas for outlining the phases of the spiritual life. St. Teresa of Avila speaks of seven mansions in an interior castle. St. John of the Cross writes about the dark nights of sense and spirit. St. Benedict teaches twelve stages up the ladder of humility. Ruth Burrows, a contemporary English Carmelite nun, writes about crossing the bridge between three islands. Classical ascetical-mystical theology emphasizes the Three Ways: purgation, illumination, and union. Another viable framework is Gaia, Samsara, and Narnia.

Gaia

Gaia is the name chosen by the ancient Greeks for the Earth Mother goddess. The word has come into our language recently through the Gaia Hypothesis of British scientist James Lovelock: "The entire range of living matter on Earth, from whales to viruses, from oaks to algae, [can] be regarded as constituting a single living entity."[1]

This idea is startling to modern science, with its tendency to divide and subdivide and sub-subdivide for academic convenience. But it is well known—indeed automatically assumed—by most primitive peoples. It is also assumed by most Christian mystics. Lovelock's hypothesis is a modern scientific description of the orthodox Roman Catholic doctrine of the Mystical body of Christ. As Christians, we refer to the earth as the Body of Christ, rather than the body of an earth mother goddess.

Our ancient Hebrew and Christian scriptures tell us the story of our intimacy with the earth. Genesis teaches us that the Lord God formed us "out of the dust of the ground" and to this earth will return us (Gn 2:7; 3:19). Psalm 139 thanks God for fashioning us "fearfully and wonderfully," "in secret," "in the depths of the earth."

The psalm-songs are full of delight and awe over the mystery of our intimacy with the earth, our intimacy with "fire and hail, snow and mist," "mountains and all hills," "sea monsters and all depths" (Ps 148). Psalm 104, one of the most lyrical earth-praisers, sings the glory of God "robed in light as with a cloak" who "spread out the heavens like a tent cloth" and "made the moon to mark the seasons."

The prophet Daniel, in a canticle the church sings on Sundays and feast days, joins with all the "works of the Lord" to bless him:

> "Let the earth bless the Lord;
> praise and exalt him above all forever.
> Mountains and hills, bless the Lord,
> everything growing from the earth bless the
>   Lord." (Dn 3:74-76)

The last two chapters of the Book of Job call upon the chorus of the morning stars, the dwelling place of light, the storehouse of snow and treasury of hail, the lions, mountain goats, deer and wild horses, the ravens, hawks, and the majestic eagle. In an earlier passage, this book urges us to learn humbly from the earth:

> You have only to ask the cattle, for them to
>   instruct you,
>   and the birds of the sky, for them to inform
>   you.
> The creeping things of earth will give you
>   lessons,
>   and the fish of the sea provide you an
>   explanation:
> there is not one such creature but will know
>   that the hand of God has arranged things
>   like this! (Jb 12:7-10)

We all know about St. Francis of Assisi and his kinship with all life: Brother Sun and Sister Moon, Brother Wind and Sister Water, Brother Fire and Sister Earth, "our mother, who feeds us in her sovereignty and produces various fruits and colored flowers and herbs." Hildegarde of Bingen, a medieval Rhineland mystic, praises the earth as the mother of all that is natural and human, "the mother of all, for contained in her are the seeds of all." Nicholas the Frenchman, who lived as a hermit on top of Mt. Carmel in Palestine, called the creatures around him his sisters and the mountains his "conventual brethren." Indeed, this view of the natural world is traditionally Christian.

In the opening pages of *Earthy Mysticism*, William McNamara, O.C.D., founder of the Spiritual Life Institute, explains that the source of Christian mysticism is in the earth, in the world, in the flesh:

> Authentic Christian mystics are notoriously earthy. They love the earth and take good care of it. . . . They recognize how sacramental the earth is. In this they resemble the North American Indian. They enjoy the earth and find their delight in it, without being inordinately attached to it. . . . They see everything as a sign, sample or symbol of God and therefore affirm the totality of being.[2]

Calling all humans "high priests of creation," Fr. William insists that our pastoral care "cannot merely include the people in our domain but must extend to the animals, trees, flowers, parks, ponds and air." We must celebrate our "significantly profound relationship with all levels of life—animals, vegetables, and mineral as well as human—affirming and consecrating our solidarity with all creation."[3] This is not the heresy of pantheism but the orthodoxy of panentheism.

In the Gaian phase of the human adventure, then, we recognize that we are made out of the very stuff of the earth. And we say, "This is my body." That makes the Gaian phase eucharistic. We give thanks as we celebrate our oneness, since *eucharist* means "thanksgiving." The first phase of our spiritual growth focuses on the mysteries of creation and incarnation, the joyful mysteries of the Rosary.

The Gaian phase is not only eucharistic but edenistic (not hedonistic, though it can deteriorate into hedonism, if we are not vigilant). We recognize the earth as a garden, a garden of Paradise, a garden of Eden.

Samsara

Samsara is a Sanskrit word used by Hindus and Buddhists to refer to the relative, finite, phenomenal quality of the world. In the Samsara phase of our spiritual growth, we see that the earth, the flesh, the world is ephemeral: fleeting, changing, passing—ultimately dying; contingent, finite, and terribly limited. We take a deeper look at our garden of Eden and see that it is somehow spoiled, invaded, with weeds sown among the wheat. According to Jacques Maritain, the French Thomistic philosopher, our garden is a "crucified Paradise."

If we live in the garden of Eden in the Gaian phase, in Samsara we live exiled from Eden. According to Fr. William, we feel the pain of exile on earth. This is "the meantime—and there is something mean about it."[4] With St. Teresa we see life as "a night in a bad inn." Singing that ancient Catholic hymn the "*Salve Regina*," we call ourselves "poor banished children of Eve, mourning and weeping in this valley of tears." How like the sorrowful mysteries of the Rosary.

We experience Samsara in our weakness: physical, and moral, and mental. We feel it in illness, aging, and death. We know it when the sun sets or the fog rolls in, when the leaves fall from the trees in autumn, when the ocean tides wash away our sand castles, or the wind blows and erases our footprints in the sand.

What profit has humanity from all its labor at which it toils under the sun? asks the book of Ecclesiastes:

> Sheer futility. . . . A generation goes, a generation comes, yet the earth stands firm for ever. The sun rises, the sun sets; then to its place it speeds and there it rises. Southward goes the wind, then turns to the north; it turns and turns again; then back to its circling goes the wind. Into the sea go all the rivers, and yet the sea is never filled, and still to their goal the rivers go. (Eccl 1:1-9)

As Duquesne University philosopher C. D. Keyes points out, the contingency of our human condition almost crushes us. We are, he says, a flash that comes and goes. We are just a spark in history. And historical time is brief compared to biological time, which is dwarfed by geological time, which, in turn, pales to insignificance in cosmic time.

Crowfoot, a Blackfoot Indian from Alberta, Canada, expresses it more poetically: "What is life? It is the flash of a firefly in the night. It is the breath of a buffalo in the wintertime. It is the little shadow which runs across the grass and loses itself in the sunset." "We are not important," says a Taos Pueblo Indian: "Our lives are simply threads, pulling along the lasting thoughts." How like the wisdom of Isaiah this is: "You have folded up my life like a weaver who severs the last thread. My dwelling, like a shepherd's tent, is struck down and borne away from me" (Is 38:10-14,17-20). Job echoes this same lament:

> For his part, he crumbles away like rotten
> wood,

> or like a moth-eaten garment,
> A human being, born of woman,
> whose life is short and full of trouble.
> Like a flower, such a one blossoms and
>   withers,
> fleeting as a shadow, transient. (Jb 14:1-3)

Perhaps no one knows the sorrow of Samsara better than the Psalmist, who sings: "My days are vanishing like smoke" (Ps 102). Or,

> Ordinary people are a mere puff of wind,
> important people are a delusion;
> set both on the scales together,
> and they are lighter than a puff of wind.
>   (Ps 62:9)

Psalm 90 goes on at length:

> You bring human beings to the dust . . .
> A thousand years are to you
> like a yesterday which has passed,
> like a watch of the night.
> You flood them with sleep
> —in the morning they will be like growing
>   grass,
> in the morning it is blossoming and growing,
> by evening it is withered and dry. . . .
> Our lives are over like a sigh.
> The span of our life is seventy years—
> eighty for those who are strong—
> but their whole extent is anxiety and trouble,
> they are over in a moment and we are gone.
>   (Ps 90:3-10)

Despite culture, education, or religious tradition, note the similarity in all this imagery, its primordial, archetypal human quality: fading flower, passing shadow, breath, sigh, dust, dream, garment that is changed or consumed by moth, severed thread.

Black Elk, one of the greatest of the Oglala Sioux, uses the image of the withered tree. In this passage he actually takes us through all three phases of the human adventure: Gaia, Samsara, and Narnia. Note carefully the crucial role played by the tiny word *but*:

> My friend, I am going to tell you the story of all life that is holy
> and is good to tell, and of us two-leggeds sharing it with the

four-leggeds and the wings of the air and all green things, for these are children of one mother and their father is one Spirit. . . .

Now that I can see it all as from a lonely hilltop, I know it was the story of a mighty vision given to a man too weak to use it; of a holy tree that should have flourished in a people's heart with flowers and singing birds, and now is withered; of a people's dream that died in the bloody snow.

But if the vision was true and mighty, as I know, the vision is true and mighty yet: for such things are of the spirit, and it is in the darkness of their eyes that men get lost.

Grandfather, Great Spirit, you have been always, and before you no one has been . . . you where the sun shines continually, whence come the day-break star and day, behold me! You where the summer lives, behold me! You in the depths of the heavens, an eagle of power, behold!

Isn't it fascinating that the original sin stories of both the Hebrew-Christian tradition and Sioux tradition revolve around a tree? Black Elk gives us the Sioux version of the Catholic "glory be to the Father . . . as it was in the beginning, is now, and ever shall be, world without end." He says: "Grandfather, Great Spirit, you have been always, and before you no one has been."

As in Black Elk's song, note the crucial use of the words *but* or *yet* in these scriptural passages, which describe the transition from Samsara into the Narnian realm:

> The whole world, for you, can no more than
>   tip a balance,
> like a drop of morning dew falling on the
>   ground.
> *Yet* you are merciful to all, because you are
>   almighty. . . .
> You love everything that exists. . . .
> You spare all, since all is yours, Lord, lover of
>   life! (Wis 11:22-26)

> As tenderly as a father treats his children,
> so Yahweh treats those who fear him;
> he knows of what we are made,
> he remembers that we are dust.
> As for a human person—his days are like
>   grass,
> he blooms like the wild flowers;
> as soon as the wind blows he is gone,
> never to be seen there again.

But Yahweh's faithful love for those who fear
   him
is from eternity and for ever. (Ps 103:13-17)

My days are like a fading shadow,
I am withering up like grass
But you, Yahweh, are enthroned for ever,
each generation in turn remembers you. . . .
Long ago you laid earth's foundations,
the heavens are the work of your hands.
They pass away but you remain;
they all war out like a garment,
like outworn clothes you change them;
but you never alter, and your years never end.
                    (Ps 102:11-12,25-28)

Narnia

If the Gaian phase is eucharistic and edenistic and the Samsara phase is ephemeral, the Narnian phase is eternal. It corresponds to the mystery of the resurrection and the glorious mysteries of the Rosary.

As St. Teresa teaches: All things pass, but God never changes. God alone suffices. We must move, then, from the changing, mediated, created being of the earth, into the unchanging, unmediated (immediate), uncreated being of the Godhead. As Fr. William wrote in one of his earliest works on earthy mysticism: "We must sink roots deep into the earth. These roots then become routes into other worlds. Because as Dostoyevsky said, ultimately our roots lie in other worlds."

Notice the eternity in all the previous "but" passages, the use of expressions such as "lasting days," "world without end," "from eternity to eternity," "from age to age." Adjectives and adverbs such as "imperishable," "changeless," "forever" and "always" imply infinity, no contingency, no limitations. Black Elk intends the same in his description of a realm where day breaks, the sun shines, and summer lives. It is important to remember here that eternity does not mean perpetuity but a *point*, the point where past, present, and future converge and become one. This is what we mean when we say that contemplation (or mysticism) is seeing everything against the background of eternity.

As Chief Seattle of the Dwamish tribe explains: "There is no death. Only a change of worlds." St. Paul makes the same astounding claim:

Now I am going to tell you a mystery: we are not all going to fall asleep, but we are all going to be changed, instantly, in the twinkling of an eye, when the last trumpet sounds. The trumpet is going to sound, and then the dead will be raised imperishable, and we shall be changed, because this perishable nature of ours must put on imperishability, this mortal nature must put on immortality.

And after this perishable nature has put on imperishability and this mortal nature has put on immortality, then will the words of scripture come true: *Death is swallowed up in victory. Death, where is your victory? Death, where is your sting?* (1 Cor 15:51-55)

This is the extraordinary good news of Christian revelation. We catch glimpses of its wonder in Gaia, as we watch daylight follow the darkness of night, spring follow the dead of winter, flowers and fruit follow the decomposition of the seed. But it takes the resurrection of Jesus to reveal this truth in all its majesty and glory! We call Jesus the first born from the dead. At first, his friends and disciples don't recognize him, because he is so brand new. Yet, at the same time, he is so ordinary and so familiar to them that they think he is the gardener, a fisherman on the beach, or another traveler on the road.

Apart from scripture, the best descriptions of this great realm of human experience, the mystery of life out of death, resurrection out of crucifixion, can be found in the *Chronicles of Narnia* by C. S. Lewis, a series of seven fairy tales as meaningful for adults as they are for children. The seventh Chronicle, *The Last Battle*, is the most relevant here, especially the last five chapters.[5]

Lewis uses utterly mysterious paradoxes and poetic images to describe the Narnian phase of the human adventure: "further up and further in," "farewell to shadow-lands," "more like the real thing," "the inside is larger than the outside," "world within world, Narnia within Narnia."[6]

Reflecting Black Elk's vision, after the last battle (of death), on the other side of the door, Narnia is a land of blue sky and warm daylight, a land where it still seems to be early and the morning freshness is in the air. Old folks there feel "unstiffened." Everyone looks cool and fresh, as though they had just come from bathing.[7]

Here's how Peter the High King describes his own experience of this mysterious realm: "I don't know. It reminds me of somewhere but I can't give it a name. Could it be somewhere we once stayed for a holiday when we were very, very small?"[8] Notice how he associates this realm of being with holiday time and childhood. As Jesus taught over and over again, "You cannot enter the kingdom of heaven unless you become like a child" (Mt 19:14).

This Narnia is familiar to Peter. It reminds him of somewhere. It is the same Narnia as always. And yet it's so brand new that Peter can't give it a name. "World within world," "the inside is larger than the outside." Compared to this new moment, the old Narnia seems like a dream, a copy, a shadow. Peter says "farewell to shadow-lands" because this is "more like the real thing." "Listen, Peter," explains Lord Digory:

When Aslan said you could never go back to Narnia, he meant the Narnia you were thinking of. But that was not the real Narnia. That had a beginning and an end. It was only a shadow or a copy of the real Narnia, which has always been here and always will be here: just as our own world, England and all, is only a shadow or copy of something in Aslan's real world. You need not mourn over Narnia, Lucy. All of the old Narnia that mattered, all the dear creatures, have been drawn into the real Narnia through the Door.[9]

What enormous comfort these words are! Everything that matters to us, whatever we have "lost"—our deceased loved ones; our lost cats and dogs; the homes, friends, and gardens we have had to leave behind—all restored to us in "the real Narnia through the Door," the Narnia that "was in the beginning, is now, and ever shall be, world without end!" So with Lucy, we "need not mourn" but rejoice! But we must remember that this new Narnia is different:

as different as a real thing is from a shadow or as waking life is from a dream. [Lord Digory's] voice stirred everyone like a trumpet as he spoke these words: but when he added under his breath "it's all in Plato, all in Plato: bless me, what *do* they teach them at these schools!" the older ones laughed. It was so exactly like the sort of thing they had heard him say long ago in that other world where his beard was grey instead of golden. He knew why they were laughing and joined in the laugh himself. But very quickly they all became grave again: for, as you know, there is a kind of happiness and wonder that makes you serious. It is too good to waste on jokes. . . . The difference between the old Narnia and the new Narnia was like this: the new one was a deeper country: every rock and flower and blade of grass looked as if it meant more. I can't describe it any better than that: if you ever get there, you will know what I mean.[10]

The magical mystical creature, the Unicorn, sums up what everyone was feeling:

I have come home at last! This is my real country! I belong here. This is the land I have been looking for all my life, though I never knew it till now. The reason why we loved the old Narnia is that it sometimes looked a little like this. Bree-hee-hee! Come further up, come further in![11]

"Home at last." Our "real country" where we belong. That "other world" where our true roots lie.

As Lucy, King Peter, their brother Edmund, the Unicorn and others keep moving further up and further in, they see a great bank of brightly colored cloud. Looking harder, they see that it is not a cloud at all but a real land:

"Why!" exclaimed Peter. "It's England. And that's the house . . . in the country where all our adventures began!" "I thought the house had been destroyed," said Edmund. "So it was," said the Faun. "But you are now looking at the England within England, the real England just as this is the real Narnia. And in that inner England no good thing is destroyed."[12]

No good thing is destroyed! As Jesus promised: "If I be lifted up, I will draw all things unto myself." St. Paul said it his own way: "And after this perishable nature has put on imperishability and this mortal nature has put on immortality . . . death is swallowed up in victory" (1 Cor 15:53-55).

As the Buddhists teach, Samsara *is* Nirvana. In other words, Gaia *is* Narnia. Heaven begins on earth. The Kingdom of God is in our midst. If we are eternal, we are eternal *now*. We can experience Narnia *now*, the fullness of life *before* death, not merely after death. We touch the eternal in time, here and now, when we live fully on the spot where we are, which is another description of the contemplative way of life.

We must sink our roots deep in Gaia. These roots become routes through Samsara into Narnian worlds. And ultimately, "our roots lie in other worlds." Paraphrasing the Unicorn's enthusiastic insight: the reason we love Gaia so much is that sometimes it looks a little like Narnia. This is what T. S. Eliot meant when he said:

> We shall not cease from exploration
> And the end of all our exploring
> Will be to arrive where we started
> And know the place for the first time.[13]

In the *Chronicles of Narnia* the central figure is Aslan the Lion, who is the Christ. Aslan describes the high reaches of the Narnian experience like this: "The term is over: the holidays have begun.

The dream is ended: this is the morning."[14] The morning. No wonder we Christians speak of Jesus as the Radiant Dawn and say that our deceased ones wake up in Christ and dwell in a realm of perpetual light!

Lewis concludes his *Chronicles* with a soaring uplifting passage about Aslan and the Greatest Mystery, which no one can describe:

> And as He spoke He no longer looked to them like a lion; but the things that began to happen after that were so great and beautiful that I cannot write them. And for us this is the end of all the stories, and we can most truly say that they all lived happily ever after. But for them it was only the beginning of the real story. All their life in this world and all their adventures in Narnia had only been the cover and the title page: now at last they were beginning Chapter One of the Great Story, which no one on earth has read: which goes on forever: in which every chapter is better than the one before.[15]

Notes

1. See James Lovelock, "Earth Mother Gaia," *Desert Call* (Summer 1983).
2. William McNamara, O.C.D., *Earthy Mysticism* (New York: Crossroad, 1983), pp. ix-x.
3. McNamara, p. 103.
4. McNamara, pp. ix-x.
5. C. S. Lewis, *The Last Battle* (New York: Collier Books, 1986 [copyright 1959]).
6. Lewis, pp. 140, 169, 180.
7. Lewis, pp. 157, 167, 138, 133.
8. Lewis, p. 167.
9. Lewis, p. 169.
10. Lewis, pp. 169-71.
11. Lewis, p. 171.
12. Lewis, p. 171.
13. T. S. Eliot, "Little Gidding," *Four Quartets*.
14. Lewis, p. 183.
15. Lewis, pp. 183-84.

15

Verbal Pollution

There can be no healing of the earth, no reform of state or church, no restoration of the unity and felicity of family, no beauty in the arts or entertainment industry, no reality in the media, no truly liberal education, no city civility, no mindful work, no renewal of humanity unless there is a radical renewal of language. The most corrupting influence in the world today—and the most tolerated— is verbal pollution, a pollution begun in childhood and fostered at home, on the streets, in school, in the marketplace, in Congress, in the White House, in churches, sports, and even between lovers. The most gruesomely grotesque feature of our cultural decline is the ugly shambles of our language.

The most blatant form of this cultural malaise is strident street language, by no means confined to the streets. It is no surprise to hear it, as if it were an acceptable style of communication, among the pious and the politically correct. By street talk I mean hollow words devoid of meaning or words that are vain, violent, and vulgar. I am not referring to simple, straightforward exchanges or earthy comments among street people but to the daily twaddle that repudiates the sacred Word, the Logos.

Nothing creates a climate of culture so much as the words we use. Build a beautiful new home, school, or office, fill it with bad language, and the place is ruined. Send a young person to the most expensive and prestigious university in the country, but no education will be possible if the student is inundated by bad language. I have not met a student in years, anywhere, who has not been lacerated by linguistic atrocities, shackled passively and respectably by the bland barbarism of academic verbosity and college crudity. The university climate, Catholic or not, is universally smarmy. But such garrulous humbug is by no means limited to the halls of academe. It's just that one would expect word vendors to be eloquent and the educative atmosphere they create to be crackling with truth.

In *The Humiliation of the Word*, French lay theologian Jacques Ellul goes so far as to distinguish between the truth and reality. This "everyday reality," for which we stand, he calls a lie. It is the Big Lie of technological society. The technique of schools, politics, social compulsions, the military-industrial complex, and even religion is to fashion and foster the Lie: the pseudo-events, the images, the fantasies, the deception people want rather than the terrible beauty, the naked truth. With our deranged language we have turned the world upside-down. There are no absolutes, no permanent values, no public philosophy, no reliable, traditional stepping stones into Ultimate Reality. We suffer this essential deprivation because we have no primordial words and have profaned our sacred words. They are now trivial and trite, used to blemish and blaspheme. Most talk is small talk, that is, chirpy chitchat and specious prattle. The rest is acrimonious invective or unctuous flattery: the tool of propaganda.

What we regard as reality is in fact a subtle sham, which we might aptly call the Empire as opposed to the real world, a modern Tower of Babel that we have constructed by gigantic heaps of words, words that enslave us by their mendacity, manipulation, and mediocrity.

Our boredom mounts. We do and say so much, but just below the superficial flourish, the windbag chatter, is a terrible tedium. Loose words, uprooted, unconnected with the original Word cause confusion. We choose the wrong revolution. We give ourselves over to the wrong prayer. We idolize the city—or is it the suburbs? We adulate gurus and therapists while we criticize and crucify priests. We prize our own individual experiences while we ignore or even deny the one mystical experience at the heart of things—the prayer of Christ, the prayer of the church. We substitute facile beliefs for a vital faith. We join the wrong movements. Most movements today, some flourishing in the Catholic church, are spurious. They come and go, their banality inevitably short-lived. Their words, cliches, slogans, and disguised idolatries sputter and spit, generating hot air and ashes instead of salutary action. Such power-fired, self-righteous eruptions are as nugatory as they are trendy and tawdry.

Shoddy Politicisms

Politicians seem to have a most pernicious effect on language. We were shocked by President Nixon's dirty tricks and dirty talk.

But, sad to say, no one seemed upset when President George Bush used spectacularly shabby language to announce to the nation our engagement in the Gulf War. "We're going over there to kick ass," said he. What an embarrassment!

Shoddy language both leads to war and is an effect of war. Madeleine L'Engle writes:

> We cannot Name or be Named without language. If our vocabulary dwindles to a few shopworn words, we are setting ourselves up for takeover by a dictator. When language becomes exhausted, our freedom dwindles—we cannot think, we do not recognize danger, injustice strikes us as no more than "the way things are."

I might even go the extreme of declaring that the deliberate diminution of vocabulary by a dictator, or an advertising copy writer, is anti-Christian.

That is why politicians are responsible for the decadence of the language. Politicians invented some of our worst euphemisms to palliate or deceive us. We have no more prisons, just correctional facilities; no more poverty, just low income levels; no more poor children, just culturally disadvantaged ones; no more destruction by bombs, just pacification; no more military invasions, just protective reaction; no more killing civilians, just collateral damage; no more old people, just senior citizens; no more lazy students, just underachievers.

Word pollution makes bureaucratic bungling seem harmless, renders anti-social behavior acceptable, and erases unpleasant images of poverty and social injustice, thus lulling people into apathy and giving war an innocent look. It has been a long time since a national shudder has been detected in the face of America's techno-barbaric juggernaut.

Psychobabble, Gossip, and Victim Language

We live in an environment whose principal product is garbage. We garble and ruin words, such as *actually, virtually, meaningful, awesome, frankly, devil's advocate, at this point in time, there for me* (or not), *yuh know, hopefully, okay, like, hey, I know where you're coming from*. Pointless conversations and the limitless proliferation of words poison our cities and our minds simultaneously, for the sake of advertising and propaganda.

Psychobabble must be distinguished from genuine psychological language—so terse and true that it doesn't hide but unveils reality. Psychobabble, in contrast, is a language unfit for the quest of holiness or even wholeness, but instead, a pitiable kind of high I.Q. whimpering. In *Sophie's Choice* William Styron catches in a poignant scene the sophisticated wretchedness and the pious pretensions of this therapeutic age. Sophie and her chums are sitting languidly by the sea. The chums are talking about their various forms of therapy, a business referred to by Walker Percy, the late great Catholic writer, as "the long suck of self." Annoyed by Sophie's silence, they ask her, "What's your problem?" Sophie, who had escaped from a German concentration camp, said she had no time for nebulous drivel over unearned unhappiness. What a perceptive commentary on our psychological age! What Sophie implies is a clear direction into both sanity and sanctity. It is an echo of the mystical teachers such as St. John of the Cross and St. Teresa of Avila.

A few years ago renowned men and women in the worldwide field of psychotherapy gathered in Phoenix, Arizona, for a convention. They came to an extraordinary conclusion: There was, in fact, insufficient evidence to indicate that their professional work was very often very helpful. There was evidence that in many cases more harm than good was done. I have before me the names of twenty-five friends who have been in therapy for years. Only one has been helped. The others have been damaged. My generation tells horror stories about confessors who hurt them. Will the next generation accuse its therapists of slick adjustments? Something like that is bound to happen. Just as there are good and bad confessors, so there are good and bad therapists.

Gossip can be a pernicious form of verbal pollution. Some of the priest-bashing that occupies a great deal of media attention is nothing but gossip. This, too, is part of the Big Lie. Many of the convicted men are clearly guilty and must be held responsible for the suffering they have created. But a good number of the accused and then publicly disgraced men—priests, brothers, even a few bishops—are innocent, cut down by neurotic men and women and a lunatic media. Here again, the church itself fails to the extent that it buys into the Big Lie instead of doing what the gospel demands of the followers of Christ—to live the truth.

We live in an age of victims, so our language is becoming a tiresome, weary language of victims. While many truly heartbroken victims remain silent, an endless chain of victims trudges noisily into our consciousness, uttering groans and grunts quite incommensurable with their burdens. *Discrimination*, a key word in education, is now known only in its pejorative sense. Seminal words like this get skewed from misuse or overuse.

Inadequate Academic and Theological Language

Abstraction—the inadequacy of words to life— is one of the worst traps of academe. This is certainly true of our seminaries. As the great contemporary theologian Dorothy Soelle says, What is called scientific theology is normally conveyed in a language devoid of a sense of awareness. It is unaware of the emotions, insensitive to what people experience. It has no interest and no appeal. It has a dull flatness because it leaves no room for doubt, that shadow of the faith. . . . Any theology that wants to communicate with real people must however use a language that shows awareness, brings them and their problems into the dialogue, and is forceful. This grows from practical experience and leads to a change in being and behavior.

Soelle highlights a false academic theological language that has three appalling results: the neglect of emotions, feelings, and experience; the absence of doubt; and the loss of dialogue. Only practical experience, closely scrutinized, will change this deplorable situation.

Theologians must speak beyond the academy. The Divinity Schools of American universities are ghettoes—respectable places, but parochial, insular, full of half-dead people speaking an almost dead language. Our society needs contemplatives and prophets who will speak the tragic, comic, mystic truth to people dehumanized on a raped earth, in a ravaged culture. Only with olympian optimism could we expect a contemplative mind, a prophetic voice, a Christian personality, to emerge from our seminaries—from either the conservatively stuffy ones or the liberally sappy ones.

What is required of us? Shakespeare knew:

> The weight of this sad time we must obey
> Speak what we feel, not what we ought to
> say.

As Evagrius of Pontus, one of our most ancient and influential Catholic teachers, said: "The theologian is one whose prayer is true." What is spoken in God's name is the Word. Otherwise we blaspheme. St. John of the Cross, whose poetic truth surpasses even Dante and Shakespeare, said that from all eternity God speaks one Word. The Word reaches its fullness in Christ. Nothing remains to be said. But cowards and bumpkins that we are, we have hardly begun to penetrate that Word. We need people of the Word, men and women of prayer, Christ-men and Christ-women. Asking clergy to be semi-professional therapists or social workers with a mildly spiritual aura is to completely misunderstand their vocation.

St. Paul assures us in his second letter to the Corinthians that God has entrusted to us his own Word (*logos*: "message") and the ministry (*diakonia*) of reconciliation. The acceptance and fulfillment of both responsibilities depend on many other significant words, for instance, all the gospels and St. Paul's letter to the Ephesians. *Reconciliation* is one of those tricky words, sometimes used as weasel words by ecclesiastical sleuths who steal from the Empire. If we skip all the trouble mentioned in Ephesians, breaking down walls of partition and being freed from slavery, if we don't risk our lives in apocalyptic warfare and go out into the desert and face the dark night of the soul—our own, and the night of the church and state—if we do not take the longest stride of soul humankind ever took, then there is no reconciliation. Unless we are inspired by and stand under the judgment of the Word—the Ultimate Truth— we will end up thinking what memorable social reformer Saul Alinsky said we would: "Reconciliation means I am in power and you get reconciled to it."

To speak soothing words that trivialize the pain of despair is an act of great cruelty and contempt to an oppressed and depressed society, to a culture in ruins. We need humble words, honest words, clothed in splendor—the shining truth—to humanize human beings.

Bad words, sloppy, careless, merely modern words can spoil even the sublime vocation of a spiritual director, a soul-friend, whose one purpose is to help others into divine union, into the love of God and the whole creation. There is nothing individualistic about this. Care of the soul and care of the earth unfold simultaneously and interactively. This unique and dramatic event is hampered when the counseling model takes over. Pampering or normalizing the encapsulated ego is something entirely different from spiritual direction. The Word does not prevail; pat words and answers do. Therapy couples and groups can be the most isolated in the world.

The Incarnate Word

God speaks. Jesus is the Word:

> The Word was with God
> and the Word was God. (Jn1:1)

From beginning to end, the Bible deals only with the Word. This Word is both act and mystery. The Word commands with absolute authority. We respond, "I am commanded, therefore I am." The Word spoken by God from the beginning is incarnate in Jesus. We know that Jesus listened so perfectly to what God said that he responded fully and so was indeed the Word. Jesus bears in himself the Word of God, so that he can say, "I am Truth and Life" (Jn 14:6), which means "I incarnate the creative word"; "I am the Way" means that the Word is the guide.

"God speaks" means that the question of truth—sheer reality—has been raised. The question of falsehood and the pollution of language is raised in that same moment of truth. The "anti-God" is named the Liar. Is this why we cannot bear to face our own modern conspiracy in the cultural lie, why we debunk our heroes and adulate our anti-heroes?

To pollute language is to falsify the Word. The vocation of the church is to contemplate the Word, conserve it, penetrate it freshly every day, and then communicate it. "By your words you will be justified, and by your words condemned" (Mt 12:37). This reminds us of the ultimate seriousness of words. Words involve commitment and reflect authenticity. We will answer "for every unfounded word" (Mt 12:36). The Word is the means God chooses to express himself. Thus the gravity of words. If we speak the truth in a world of acceptable lies we, as witnesses of pure veracity, become martyrs.

Pope John Paul II, a living witness to immutable truth—branded in the nature of things and revealed in tradition and the Bible—is, ironically, the world's favorite martyr. The world (more correctly, the Empire, the fallen world, the popular fabrication of pseudo-events, false values, and half-truths) loves him and kills him, or simply ignores him, the speaker of the truth, a contemporary hero who is trying to restore to the human word its fullness, its refined ferocity. Like all great heroes, he is overwhelmed, broken and crushed by the truth of this word he must speak. Shades of Augustine, Joan of Arc, and Kierkegaard.

Preaching is the most frightful adventure. If the preacher is not inspirited by the Ineffable, if the words spoken are not born in silence, in eternity, then the words become very subtly and insidiously the Lie that fills the sound and fury of one aspect of this age that is dying of hunger for the truth.

What is required of the church? Fidelity to the Word. It must become detached from every socio-cultural milieu and remain permanently and robustly attached to the living Word, to the ongoing command of God who speaks at the heart of things.

We worry when the church loses or abandons imperial power for the sake of love. We suspect anarchy when the church refuses to be political except as lover, as Bride of Christ. But it's either poverty or relying on power. Poverty is an evangelical imperative. If the church is powerless—like Christ—it has no power to rely on. God

is love. Period. Jesus embodied that love. So does the church—
more or less. The glory of that love must not be mistaken for a
power play. In a shrunken age it may seem spectacular, but it is no
spectacle—just the splendor of the Word exploding amid the egre-
gious folly of our wooden, wastrel words.

Reverence toward Language

The truth is simple: God has spoken. The Word was made flesh.
Who will be faithful to his flesh? Only heroes. *Orthodoxy* literally
means "right glory." To enjoy right glory we need the right words.
Only contemplatives to whom the eternal Word is revealed can speak
with humble and unswerving conviction, with personal and unwa-
vering authority. Contemplatives do not represent the crowd but
rather the Poor Man, Jesus Christ. People who penetrate and per-
petuate the Word are speakers of the Truth and witnesses of
Ultimate Reality in such a remarkable way that their orthodoxy
may be mistaken for anarchy. In fact, the more God-centered and
obedient they are to the Word, the more powerless they will be, and
only as such are they spiritually authentic and politically significant. The rest of us are frauds.

Part of the priest's vocation is to create a sacred center for society. Toward this end he must save words from corruption and purify
the language. No one suffers more from polluted language than
the priest. That is why he must make a stupendous effort to preserve and present a pure mind and heart. Bombarded by so much
pietistic poison, he needs to be heroic, his language purely eloquent. Dag Hammarskjold, former secretary-general of the United
Nations—and a good example of a contemplative in action—thought
that good language was a moral imperative for priests, politicians,
and the proletariat. "To misuse a word," he said, "is to show contempt for Man. It undermines the bridges and poisons the wells. It
causes Man to regress down the long path of evolution."

Everyone needs to approach language with reverence. As Dylan
Thomas said, "I write for the glory of God and I'd be a damned fool
if I didn't." Exactly! Can you imagine, for the glory of God, writing
gobbledygook?

Let me now be positive about language. According to linguistics
scholar Peter Farb, in *Word Play*: "Something happened in evolution to create Man the talker." I'd say God created us to participate
in the eternal Word. We are talkers. Speech is our most exalting
tool, the toy, the comfort, and joy of the human species. The pity is
that talkers often go so far beyond the line of what is needed and
desired that they are listened to with either boredom or contempt.

Whenever language is good, it is luminously gravid with truth. Professors Arn and Charlene Tibbitts of the University of Illinois say it all in the pithy paragraph that concludes their book *What's Happening to American English*:

> As our public rhetoric becomes more churlish and illogical; as our plays, movies, and works of fiction become more slovenly, violent, and dirty; as our scholarship becomes more pretentious and inaccurate, so our American spirit loses its magnanimity and grows weak and mean-spirited. . . . To abuse language is to abuse the very idea of being human. Wit, kindness, intelligence, grace, humor, love, honor—all require the right employment of language for their embodiment. In the decline of American English may also be seen the decline of the American as human being.

The Art of Conversation

On every level of society, whatever our way of life, we have lost the art of conversation—good talk. Without this kind of intimate relatedness in love and leisure, with no utility, not even high purposes in mind, our portion of the earth will be endangered. The focused energy of the nation must be spent on this urgent endeavor: to converse, to engage in rational discourse, to speak with such eloquent logic and enlightened love that words are born in a deep, profound silence, expressed with reverence, clothed in splendor, drawing all those who dare to speak—to break the pure mystery of silence—into an act of holy communion.

On this subject I can make an experiential comment. I live with two communities of monks, one in Crestone, Colorado, one in Kemptville, Nova Scotia. We have worked on the art of conversation for thirty-three years. The matrix of our talk is hours of silence and solitude. The vocal practice happens at the two meals we share together with our retreatants each week. During those meals, with as many as fifteen people at one table, we have achieved a stunning victory, and that is one conversation. Fantastic! But it will take as many years as we have left, though we are a young community, to enjoy consistent good quality. It's up and down now—sometimes dialogic ecstasy, other times, the pits. Last Sunday, for instance, we went from an exalted experience of community prayer into a breakfast conversation that was inflated, uninspired, downright boring, and worst of all, totally unconnected with the peak experience we had just shared and the feast day we were celebrating. Our lives as monks—shepherds of being who savor the Word and therefore speak with gusto muted by awe, wonder, and wise patience—were momentarily defeated. But the struggle goes on with deep commitment.

Fortunately, the struggle continues in an oasis of verbal purity and delight. In thirty-three years I have never heard a bad word uttered by one of our monks. We are not only revolutionary because our Crestone center, as far as we know, is the only solar-heated monastery in the world. In a far more crucial ecological revolution we work toward a cleaner planetary environment by creating a cleaner verbal atmosphere, freed from the pollution of dirty talk.

The ecological question is: How can we clean up the earth? More specifically, how can we acknowledge and preserve the inherent goodness of animals, vegetation, air, water, soil, and silence? The supreme ecological question is: How can we clean up our language? It can and must be done.

If certain base words have crept into our vocabulary and seem impossible to expel, then why not transform them? Forty percent of Boston's population is illiterate. And Boston is more literate than New York or Los Angeles. So our effort to clean up must be simple. Ugly words dominate our street talk and movie language. If these words cannot be eliminated, they can certainly be transformed. The devil is damned. We can say so. Hell is hell. When used verbally and meant as such, it's a good word. If we make a hell of our lives, we will probably end up in hell. That's a meaningful sentence.

We suffer from a final problem, the biggest and most terrible problem of all—our language is obliterated, the conducive climate devastated, the possibility of education made utterly impossible when the name of God is taken in vain. This is sinful. When we do it, we have no access to the Word. Speech self-destructs. It is probably the most serious sin most of us commit on a daily basis, mucking up the whole atmosphere and tempting the wrath of God. We take his name in vain both piously and profanely.

Whenever we casually refer to God or Christ or Jesus, using these names without fear, love, awe, and adoration, we are in peril. The holy name of God is the sacred center of the Hebrew and Christian traditions, the heart and soul of Western culture. God's name, Our Father, shatters our autonomy, our original sin, which is also our modern sin: the egotistic attempt to name things, to decide what is right and what is wrong, what is good and what is bad. We bow before the Infinite and at the sound of his name. And though wholly Other and transcendent, we know God is not separate. So we bow with wonder and radical astonishment before all creatures, the things God found so good in the act of creation. This creative act of God keeps things from falling apart. So do we, insofar as we are co-creators, alive to God and therefore responsible stewards of the creation. Everything matters. That doesn't mean that anything goes. It means doing what is right and saying what is true, so that "all things shall be well, all manner of things shall be well." Julian of Norwich, fourteenth-century English mystic, knew this secret when she discovered God in a hazelnut.

As Brian Moore, author of *The Catholics*, says, "When our words become prayer, God comes."

16

An Eco-Prophetic Parish?

PAULA GONZÁLEZ, S.C.

The question with which I begin this chapter is whether ecology can become the subject of a spiritual awakening in parishes.

A New Understanding of Earth and Humanity

The famous photos of earth that we received as a Christmas present from NASA some twenty-five years ago have affected our vision of earth and humanity's place in the cosmos profoundly. To understand just how much of an effect they have had, go to your attic or to the public library. Dig out a magazine or newspaper from 1969. Go through it carefully and count the occurrences of the words *global* and *planetary*. You will probably not find them at all. Yet in 1994 most of us have internalized these photographs and are beginning to understand ourselves as global or planetary citizens on a small planet in the midst of an immense cosmos.

A spirit of genuine love and concern for our planetary home is being inspired by insights from both science and spirituality. We are invited to reenvision the role of the human family in the story of the universe, in order that we may ensure continuation of humanity as part of creation's long story. Yet all is not well. If we read the signs of the times correctly, we find that we are the first generation in history that cannot assume that human life will continue. According to many competent ecologists, unless the industrial assault on earth is reduced drastically in the very near future, this

---

A longer version of this chapter was published as a series of articles appearing between November 1990 and March 1993 in *Today's Parish*. The editors are grateful for permission to abridge and edit them for this volume.

planet may not be able to support human life by the late twenty-first century.

The Need for Conversion

Time will tell if popular concern will go deeper than the level of fads. For us in the most natural community unit of our Christian life—the parish—there is, I believe, need for a spiritual reawakening that includes the earth. For Christians the so-called environmental crisis may well be a blessing in disguise. It calls each of us to be converted from consumerism to what I sometimes call the "six C's"— connectedness, collaboration, creativity, community, commitment, and celebration. The environmental situation we face calls us to an embodied faith, just as our parishes are a potentially potent social locus wherein we can be schooled in a new consciousness of our status as children of God and the cosmos.

I invite you to consider the six C's through which we are invited to experience the freedom of the children of God. In a very real way we can use everyday news items about air and water pollution, toxic waste, overflowing landfills, endangered species, and global warming as channels of God's grace. They are alarm clocks reminding us of what our throwaway society is doing to the primal air, water, earth, and fire that the Eternal Creator—the Divine Architect—made the basis of all life.

Perhaps even more poignantly they remind us how far we have wandered from realizing that *everything* is sacred. As urban, industrial societies, we have a very different blueprint—one totally out of synch with the reality of this finite planet's design. In nature everything is connected to everything else. Thus interdependence is the rule. Ecologically speaking, *community* is the fundamental pattern of all ecosystems—deserts, mountains, oceans, forests, rivers. Thus everywhere that life exists there is recycling—earth, air, and water being energized into living creatures by the fire of a star.

Today we are called again to follow a star—our own star, the sun—in much the way the Magi followed theirs at the first Christmas. As the source of life on our tiny earth, the sun directs us to a new understanding of the Christmas story. Christ is born anew in every age, but only to new "wise ones" who follow the age's mysterious star.

Creation as Process

Over the past several generations, we humans have been gifted by science with a vision that helps us understand ourselves in an immense cosmic time span billions of years old in age and billions of light years in extension. New light on the age of earth shines forth from this reconceptualization of the universe. Although theologians and other spiritual thinkers have spoken in terms of an event that occurred "in the beginning," many of us have a sense that creation is a completed action, and that now God is resting. Nothing could be further from the truth.

As in so many other contemporary areas, where we are moving from concentration on events to considering process, we are today invited to see creation as a continuing process. It may have begun, physicists and astronomers say, as a stupendous energy event, which we call the Big Bang. This giant fireworks display has been going on for fifteen billion years, and its present expression is the world we see all about us.

Thomas Berry, a priest and member of the Passionist community, calls himself a "geologian." Berry is perhaps today's best spokesperson for this new vision of creation as process. He calls it the Universe Story and invites us to reenvision humanity's place in God's plan. "The human being," he says, "is that being in whom the earth becomes conscious."

If Eternal Mystery has been celebrating Christmas—the Word made universe-flesh—for fifteen billion years, we can begin to comprehend in a whole new way what it means to be part of the Mystical Body of Christ. For through these fresh insights from science and spirituality, we are enabled to welcome the Cosmic Christ of whom St. Paul and St. John wrote so often:

> In the beginning was the Word;
> the Word was with God
> and the Word was God. . . .
> Through him all things came into being,
> not one thing came into being except through
> him. (Jn 1:1-3)

Change the tense of the verbs in these statements, and you can hear God's Word in everything you experience. The Eternal Artist speaks, sings, dances, energizes our times—and thus makes everything sacred. The very process of cosmic life is an embodiment of God's Word.

Healing Earth

A Cosmic Epiphany

In the Hispanic and Orthodox worlds, Epiphany is celebrated with more fanfare than Christmas. Epiphany, in those traditions, commemorates the moment when the Christmas mystery began to become a worldwide revelation. The Wise Men had traveled a long way before they arrived at the destination to which the star led them. They were not kings, but scripture portrays them as persons of importance, perhaps akin to world-renowned scientists in our own day. When they found that the one to whom prophecies led them was an infant, they did not question this unexpected finding.

Today we are invited to explore new and perhaps unexpected lights that shine in our weary world. Insights from science and spirituality call us to reenvision through faith-filled eyes the role we humans are supposed to play in God's universe. This may be expressed as an invitation to experience life as a "cosmic epiphany." Some people find a shift from our traditional human-centered (homocentric) understanding of the universe to a biocentric orientation a "demotion" for humankind. And perhaps it would be if the living world (the biosphere) were organized hierarchically. But it isn't.

The wonderful phenomenon we call life actually mirrors the Divine Architect's very nature. In other words, all creation is "made in the image and likeness of God," and the Trinitarian nature of the Eternal Mystery is strikingly clear.

Gimmick ecology will save no one. For Christians, as for all peoples, healing earth must be tied to deep spiritual awakening. Changes in practice will endure only if tied to changes in thinking.

Earth as Gaia

First Things First

One of the most exciting ideas to introduce a change in thinking in parishes is the Gaia Hypothesis. But if this central insight is to make sense, we need to back it up and begin with a more basic concept—that of the ecosystem, the underlying design of all living systems.

Ordinary parishioners who have not taken a special interest in ecological matters may not be familiar with the term *ecosystem*, but even tiny children can distinguish between an ocean, forest, mountain, and desert. On land we immediately identify certain plants and animals, a particular type of soil, the humidity level (related to the annual rainfall), and the average temperature (related to amounts of sunshine and cloud cover). Automatically we find ourselves making an inventory of the conditions for life that exist between the earth, air, water, and fire. Both the number and variety of living creatures are related directly to how difficult it is for life processes to be maintained. Thus, it is not surprising to find a limited variety of life forms in the arctic and an almost unbelievable profusion in the tropics.

Communities

Even this brief tour of ecosystems can help us remember that living creatures never exist alone (as we tend to keep them in today's industrialized agricultural practices). Integral to every type of ecosystem are three different categories of creatures: 1) *producers*, our green brothers and sisters, who capture the energy of the sun; 2) *consumers*, mostly animals (including ourselves), who are totally dependent on producers; and 3) *decomposers*, the tiny (and we usually think of them as lowly) creatures in the soil and oceans that liberate the atoms and molecules of the universe so that they can be used over and over again.

These three groups interact intimately with one another and with the physical environment. It is interesting to note that the general name ecologists have given to these three groups of critters is community. An ecosystem is made up of interacting, interdependent communities of living creatures. This pattern, which is fundamental in the universe, also can help us understand the nature of our God, who is the source of life. For the Trinity is actually a community so intimately interactive and interdependent as to be a single Mystery. Thus, life on this planet images God very accurately. It is this stupendous phenomenon we call life that is central to the universe story.

Changing the Model

The human chapter in this story is the equivalent to the last second of the twenty-four hours since life began on our planet. We might be tempted to think that this means our arrival is unimportant, for we are accustomed to a hierarchical understanding of relationships, with higher members as the most important. The major design principle of the model we are accustomed to using is that of a triangle standing upright on its base. As we are beginning to realize, there is something wrong with this picture.

What have all the liberation movements in human history been trying to do? Change the model! But it is only in our times—through the communications and transportation technologies that human genius has developed—that we are beginning to understand that this deep intuition, this imaging of the divine, is what is driving the urgent human desire to live in the freedom of the children of God.

Nature

Where might we learn a new model that would ensure true freedom for all members of the earth community? By turning to nature and by seeing ourselves as part of it. Here we find a deeper understanding of what the Divine Architect is calling us to become.

In the living world one pattern is repeated again and again, and its shape is roundness. Cycling is the key to life in the divine design, so the circle is a very successful model that we might strive to use in designing new human institutions—one that is horizontal rather than vertical (like the upright triangle we have been using). Imagine what might happen if human beings succeeded in toppling the triangle and removing the sharp edges?

Actually, we already know how effective this can be. After decades of technological invention, we humans have designed a wonderful social invention that follows nature's pattern and promises to revolutionize human affairs. This is the support group.

What dramatically distinguishes these from bureaucratic groups is not the absence of leaders and experts, but the way everyone in the group plays both these roles. In other words, the talents and wisdom of every person contribute to the effectiveness of the group.

This is a profoundly organic model. Think of how your own body functions. Each of the many different kinds of cells that compose you—muscle, liver, blood, brain, kidney, lung—contributes to your healthy state by doing its specific job well and interacting harmoniously with all other cell types, each of which performs its unique function.

Earth, a Living Creature

With these ideas as background, we are now ready to consider one of the fascinating new scientific insights that is just beginning to be explored—the Gaia Hypothesis. For the ancient Greeks, many finite things in the universe were a god or goddess. Planet Earth was the goddess Gaia or Ge (geology, geography, and so forth come from this root). This was also the name chosen by James Lovelock, a British scientist and philosopher, for the theory he proposed in his 1979 book *Gaia: A New Look at Life on Earth.* Lovelock and Lynn Margulis, an American biologist who had been working with him as a NASA consultant, arrived at a truly astounding realization. The earth—in its totality—exhibits characteristics associated only with living systems. Could it be, they suggested, that our little planet is a single living creature?

What does it mean to say, "It behaves as if it were alive?" Our earth has maintained a steady-state temperature for the past three-and-a-half billion years, though it should have gotten progressively hotter. The oceans should have been getting progressively saltier during this time, too, but they have not.

Perhaps most amazing of all is the composition of the atmosphere and what has happened to it in the time since life emerged. Earth and our neighbors Venus and Mars were very similar in chemical makeup when they formed four-and-a-half billion years ago. Thus the atmospheres that surrounded them would have been very similar—nearly all carbon dioxide, with only traces of oxygen and nitrogen. So, how can we explain earth's atmosphere today—this gaseous mixture that permits life to exist? Our air contains about 79 percent nitrogen, almost 21 percent oxygen, and 0.03 percent carbon dioxide. How could this dramatic change occur? Where did the 99+ percent carbon dioxide go?

What happened is that the primal carbon dioxide gas *became* the carbon compounds that make up all living cells. It was the progressive emergence of living creatures in the primeval oceans that changed the atmosphere to the air we know today—that marvelous gaseous envelope that can sustain life. Not only has the development of living creatures produced an atmosphere that supports life as we know it, these same plants, animals, and microbes keep the air's composition constant and thus supportive of the continued life of the biosphere.

A Living *Being?*

What might it mean if, in the Great Architect's design, our planetary home is a single living organism? How would each of us—and all the rest of God's creatures—fit into the picture? Perhaps each of us is a cell of this body, since we know that all living creatures are composed of cells.

For Christians, this new theory should not be a surprise, for we claim to believe in the Mystical Body of Christ. In this theology, so prevalent in St. Paul, all the people who ever lived, who live now, and who will come in the future, in some mysterious way make up the Body of Christ. In a sense, the Gaia theory simply expands the membership in the Mystical Body to include everything that has emerged from God's creative act. God's Word has been in the process of becoming the Cosmic Christ so wonderfully announced by St. John and St. Paul.

The Thinking Layer

Seen through the eyes of faith, this Gaia Hypothesis, coming from science, expresses much the same concept found in Teilhard de Chardin and, more recently, Thomas Berry. These are the first Catholic thinkers to promote a vision of the universe as an evolutionary process. Berry speaks of the Universe Story as the fifteen-billion-year "energy event" that we experience all around us. For Berry, the universe is "the primary revelation of the divine." Thus, to study and appreciate Gaia and her neighbors is to read the scripture of the universe. It is a brilliant new light inviting us to follow this star and to plumb new depths of the divine mystery.

Teilhard de Chardin was intrigued by the meaning of the emergence of the human species in the evolutionary history of the earth.

In *The Phenomenon of Man* Teilhard summarized much of this thinking in the suggestion that, with the arrival of humanity, earth got a
"new skin." Added to the primal lithosphere, atmosphere, and hy-
drosphere, and the more recent biosphere, was a "noosphere," a
thinking layer. (The Greek word *nous* means "thought" or "consciousness.") With the arrival of humanity, earth became conscious.

What does it mean for us to be the consciousness of the planet?
Recently another scientific concept has shed some light on this
question. In 1983 Peter Russell published *The Global Brain*. Much
like Teilhard de Chardin and Thomas Berry, Russell proposes that
the collectivity of humans makes up the consciousness—the brain
of Gaia. He feels that we are on the threshold of a major shift in the
manner in which the human race functions. Through today's technology the cells of the planetary brain are becoming more and more
linked, in much the same way as increasing connections form in
the brain of a human being during its development to maturity.

Russell calls us to review aspects of the story of evolution. About
ten billion different kinds of molecules had to form before the complexity of a living cell could emerge, he reminds us. In the same
way, it is known that about ten billion human brains (consciousness) are present as part of Gaia. A new phenomenon may occur
as population increases. Humans may be linked by a widespread
planetary consciousness that will result in an era of collaboration.
This dramatic change will arise from the conscious realization on
the part of increasing numbers of humans that we are meant to be
one bread, one body.

A Divine Invitation

Let us recognize these insights from science and spirituality as
bright stars leading us to God. Like the Magi, let us have the deep
faith to recognize the Divine Mystery revealing itself to us in new
and unexpected ways.

What might happen to a parish that decided to read the scripture of the universe to uncover the secrets of what it could become?
What new reverence toward our planetary home might arise in a
parish that began to interpret the ecological bad news as a divine
invitation to live in the freedom of the children of God?

In that context, Easter—the wonderful season of awakening, of
renewal, of new life—is the springtime. Every spring, earth invites
us to "become as little children," and thus experience the joy that
sparks in the eyes of little girls and boys as they see their first
butterfly or the wonder of a tulip as it opens its face to the sun.
Jesus reminds us of this in the brief but profound words he left the

apostles when they asked to learn how to pray: "Thy kingdom come
. . . on earth as it is in heaven."

What would happen to our lives if we came to realize that "come"
refers not to some moment in the dim future but to now? The delight in children's eyes results from understanding this truth. If
adults are willing to re-experience the world through the eyes of
faith, we too can experience the wonder and awe that the mystics
of all ages have proclaimed

The Mystery

Actually, the first celebration of the death-resurrection mystery,
which is the bedrock of Christian faith, did not occur only two
thousand years ago. The deep wisdom of earth has been part of the
design of the evolving universe for nearly four billion years, since
the wonder we call life began to express the Word of the Great
Architect.

Chief Seattle once said to a delegation sent from Washington to
buy land in the Northwest:

> How can one buy or sell the air, the warmth of the land? That
> is difficult for us to imagine. We are part of the earth, and the
> earth is part of us. Humankind has not woven the web of life.
> We are but one thread within it.

He also warned: "Whatever happens to the animals will happen
soon also to human beings. Continue to soil your bed and one
night you will suffocate in your own waste."

Back in Synch

The environmental crisis is an alarm calling humans to be humble
enough to learn from earth's age-old secrets and from the truly
simple people who still understand them. Life has been gloriously
successful for nearly four billion years. Yet in less than three hundred years industrial society has made changes in air, earth, and
water that threaten the existence of life as we know it.

The most immediate way to reconnect with the design of life in
the universe is to experience it. Ecosystems are complex interactions among many different animals, plants, and micro-organisms
and their earth-air-water surroundings. Production and consumption are two features of ecosystem design that have been copied in
the industrial model. Isn't it amazing that it has taken us so long
to realize that unless we incorporate decomposing, the natural resource supply may run out? In nature there are never scarce
resources, nor are there dumps, because what we call trash is
really a resource if we learn nature's most fundamental lesson:

Life comes from death. New life emerges every spring from the leaves, bark, and bones that have returned to earth and been regenerated into the simple materials from which green plants know how to make what we call food, the only source of life to us consumers.

A New Perspective for Parishes

Compost and the Pascal Mystery

Parishes could be instrumental in helping people understand Easter in a refreshing new way by having workshops on composting—teaching everyone to enter into the mystery of cycling. Many families have space in their back yards for a compost bin. Most home improvement stores have several types of bins (most made from recycled plastic). A Saturday workshop could bring folks from the parish together to begin this new adventure with a built-in support group. This could become the nucleus of outreach for a small Christian community. Members will be further enriched by sharing and celebrating the deep spiritual truths they can learn from their efforts to reconnect experientially with earth's secrets.

Social and Theological Reflection on Water

Water is life for all creatures on our tiny island planet. Surely it is not coincidental that for Christians baptismal water is the entrance to the new life that Jesus promises.

Almost all water contains chemicals that result from agricultural and industrial processes. Most of these have been added during the relatively brief 250 years of the industrial era, and virtually all these additives are harmful to living creatures.

How can we translate such facts into meditations that motivate us to "repent and sin no more"? An ecological examination of conscience can help. Every day an average North American uses 65 to 120 gallons of water—all of which has been purified through some very expensive processes. A good part of that water is used to flush away the very household and industrial chemicals that are threatening our environment. Ultimately, most of this moves into an expensive sewage system, then to streams and rivers, and finally into the oceans.

In Genesis, humankind is pictured as being given dominion over the earth, but dominion has often been misinterpreted as domination. In the Hebrew sense, however, to have dominion is to be entrusted with a sacred charge. Are we aware of our sacred charge to maintain the divine order as God has created it? In regard to water, this means keeping it free of foreign materials and using over and over again, not only water but the natural substances it carries.

What can we do? Minimally, both in our homes and in our parishes we can reduce consumption by installing devices to reduce the amount of water used to flush toilets and by putting flow constrictors on showers. We can run washers only with full loads.

A Personal Story

More basically, we can take up the challenge of what the Vatican Council II *Constitution on the Church in the Modern World* recommended we do to be convinced that we have the responsibility to provide future generations with reasons for hope and optimism. It is sometimes surprising what this can lead to. Ten years ago I began reading a good deal in so-called futurist publications about the concept of preferred futuring. I began feeling I could and should use less fossil fuel for the rest of my life, thus reducing carbon dioxide and other polluting gas emissions.

As a result of that, another sister and I pay only about thirty dollars per month in total energy costs in a lovely, super-insulated twelve-hundred square foot apartment made from an old chicken barn using volunteer help, reused materials, and new materials paid for by recycling. If parishes became centers where such things are experimented with, they could easily become centers for hope, modeling new ways of living sustainably.

17

Was St. Francis a Deep Ecologist?

KEITH WARNER, O.F.M.

On Easter Sunday 1980, Pope John Paul II proclaimed St. Francis of Assisi the patron saint of ecology, following the suggestion offered thirteen years previously by Lynn White, Jr., in his seminal article on Christianity and ecology.[1] Since then, remarkably little has been written on St. Francis's relationship with nature and even less attention has been given to exploring his environmental philosophy. What can we learn about environmental ethics from Francis? And how can the ecological wisdom of Francis promote dialogue between Christianity and ecology?

Since the advent of the contemporary environmental movement thirty years ago, different understandings of how human beings should relate to their environment have emerged. One of the most vigorous current debates occurring among environmental circles centers around a form of ecological philosophy called deep ecology. According to Arne Naess, the Norwegian philosopher who coined this term, deep ecology (as distinguished from shallow ecology, which is the usual, short-term view of nature) asks deeper questions.

> The adjective "deep" stresses that we ask why and how, where others do not. . . . We question our society's underlying assumptions. We ask which society, which education, which form of religion is beneficial for all life on the planet as a whole, and then we ask further what we need to do . . . to make the necessary changes.[2]

Deep ecology seeks to challenge our culture's fundamental human assumptions, especially those which have led us to accept materialism, militarism, and human domination over nature as normal human behaviors. Deep ecology seeks to understand and challenge the root causes of our planetary despoliation.

In the following pages I will discuss the deep ecology principles lived and preached by Francis of Assisi. By his life and writings, I hope to show, Francis demonstrated a way of living in harmony with nature that provides a much needed basis for dialogue between ecology and Christianity. I acknowledge the danger of misinterpreting Francis by bringing him to a discussion in today's vastly different cultural context,[3] but I feel this risk is outweighed by the benefit we can gain from understanding the value he placed on nature and his spiritual experiences in it. After introducing several basic concepts in deep ecology, I will discuss the ways Francis anticipated these concepts and how he differed from them. I will conclude by proposing the development of a distinctly Franciscan model of relating to Creation.

What Is Deep Ecology?

Arne Naess began developing the basic tenets of deep ecology in the late 1970s, because he was frustrated by the failure of most ecologists and scientists to address root causes of our environmental crisis. He decries shallow ecology, which studies and analyzes small sections of biological life while ignoring human behavior that threatens to destroy entire ecosystems. Naess and his followers want to ask more fundamental questions, such as, What changes do we need to make in our understanding of the world so that other forms of life can continue?

According to Naess, there are three basic principles of deep ecology. First, all life, human and nonhuman, has value in itself, independent of human purposes, and humans have no right to reduce its richness or diversity except for *vital* needs. Second, humans at present are far too numerous and intrusive with respect to other life forms and the living earth, with disastrous consequences for all, and must achieve a substantial decrease in population to permit the flourishing of human and nonhuman life. Third, to achieve this requisite balance, significant changes in human economic, technical, and ideological structures must be made. Humans must move toward stressing not bigness, growth, and higher standards of living, but rather sustainable societies emphasizing the nonmaterial quality of life. From these three basic principles other deep ecologists have identified a more precise platform, which specifies political values flowing from deep ecology: the primacy of wilderness, a sense of place, opposition to stewardship, opposition to industrial society, spirituality, and self-realization.[4]

Naess believes that almost every religious movement, from Buddhism to Christianity, has some elements consistent with deep ecology already present within it, and he challenges all his readers to identify principles of deep ecology in their respective traditions. I would like to develop some themes in the life of Francis that coincide with these principles.

Francis and Conversion

Francis Bernadone was born into a family of Italian cloth merchants around 1182. His father was an active member of the emerging bourgeois class. After several failed attempts as a knight errant, Francis went through a long period of illness during his early twenties, 1203-4, and after these experiences, his early biographers tell us, he began to undergo a profound change. He spent a lot of time by himself in the wilderness, especially in some caves near Assisi. Francis was trying to work through the beginnings of his conversion, trying to lay a foundation for living a holy life. Thomas of Celano wrote that Francis was so exhausted when he emerged from his struggles in the cave that he "seemed to be a different person than when he went in."[5]

This was a period of intense inner questioning for Francis, a time when he began to reevaluate the fundamental understanding of life which he had held up to that point in his life. He questioned the militarism, violence, and greed which were part of the culture he knew. He later became one of Christianity's most famous preachers of peace and nonviolence. Beyond preaching he expressed his conversion by his lifestyle: he moved from the comforts of the city of Assisi to the leprosarium on the periphery, he abandoned the selfishness of his youth for a life of itinerant preaching, and in his later years he spurned the popularity his fame had brought him for the life of a mystic and hermit in the wilderness. His questioning and refutation of his culture's values indicate the same probing that Naess articulated as the essentials of deep ecology. By his life and preaching, Francis challenged the basic paradigm of his culture by refusing to live by its secular values, and he encouraged his followers to do the same.

How Francis Viewed Nature

Francis valued the diversity and beauty he saw in nature. Celano wrote that Francis insisted his brothers leave a border around the community garden untouched so that wild grasses and flowers could announce the beauty of the Creator.[6] He forbade his followers to cut down a whole tree so that it might have hope of sprouting again.[7] By these concerns Francis indicated that he valued the existence of other creatures and wild plants for their own sake, that they have intrinsic value. With refreshing simplicity he calls us to refrain from a human greed that unnecessarily injures other aspects of creation.

Francis loved creation because in it he found the Lord, but his love was not limited to popular or beautiful species. Celano wrote:

> He was aflame with love for even the humble worms, because he had read that it was said of the Savior: "I am a worm, and no man" (Psalm 21:7). For that reason he would pick them up off the road and put them in some safe place so they would not be trampled underfoot by the passers-by.[8]

Francis found God in worms. In a creature not known for being powerful, quick, intelligent, or strong, he found Christ. Francis related with compassion to all creation, whether human lepers or lowly worms.

Yet Francis appreciated more than just individual species; he viewed the created world as a network of relationships, with individual elements living together in harmony joining together in an orchestra of glorious beauty.

> When he found an abundance of flowers, he preached to them and invited them to praise the Lord as though they were endowed with reason. In the same way he exhorted with the sincerest purity grainfields and vineyards, stones and forests and all the beautiful things of the fields, fountains of water and the green things of the gardens, earth and fire, air and wind, to love God and serve Him willingly.[9]

Francis was a sensuous Italian, and he experienced through his senses the vibrancy of creation around him, believing that in their very act of being, all creatures could give praise to God. By affirming the roles of and relationships among the diverse elements of nature, Francis offered a spiritual description of the interdependence of an ecosystem.

Francis and Vows

Beginning around 1210 the fraternity Francis founded began to grow exponentially, and he was no longer able to instruct new members personally. His brothers needed a constitution that would guide them, and so he wrote a Rule; in 1223 Pope Honorius III approved it. Francis adopted the monastic vows and used them to help communicate the gospel, which is what he wanted to live. The opening line of his Rule reads:

> The rule and life of the Friars Minor is this: to observe the holy Gospel of our Lord Jesus Christ by living in obedience, without anything of their own, and in chastity.[10]

In the history of Roman Catholic Orders, poverty, chastity, and obedience came to be known as the evangelical vows, and Franciscans still profess them. They guide us in our efforts to imitate Jesus Christ and to follow his gospel. I believe that each vow, interpreted today, can offer us some insight into how Franciscans can live out Francis's values in a way that is sensitive to nature.

Francis is most famous as a preacher and lover of poverty, but for him poverty was not an ideology, but rather a personal commitment, and he spoke of having a relationship with Lady Poverty. He loved her because he saw Christ's incarnation as the ultimate expression of poverty and thus the clearest expression of God's love for the world. For Francis, to imitate Christ meant to live a life of simplicity and poverty. In the same way, those who look to Francis's example can simplify their lives and live with fewer possessions—and escape the manic greed which possesses North America. Only by standing against the all-consuming avarice of our culture, which creates demand for environmental degradation around the globe, can we enter fully into the joy Francis found in creation.

In "The Canticle of the Creatures" (the complete text closes this chapter), Francis refers to "Sister Water, which is very useful and precious and chaste." Francis obviously intended chastity to refer to something other than illicit sexual activity. A broader definition of chastity can be understood to mean the right ordering of relationships, the correct kind and balance of relationships with all persons. The meaning of the Hebrew word *shalom* may cast light onto his understanding: more than simply "peace," shalom implies rightly ordered relationships that lead to the peace of God. I believe that Francis understood chastity in this context, and he saw Sister Water as a kind of example. Most living creatures are composed primarily of water, and one of our primary needs is making sure that we take in as much water as we transpire. We depend on our environment to provide water and all nonhuman life depends on us to return clean, safe water to the environment. A contemporary interpretation of chastity can expand the frame of reference for this right ordering of interdependent relationships to include all aspects of creation.

Francis was renown for his humility, and his understanding of obedience should be understood in this context. Francis insisted on placing himself and his order in the hands of the church, specifically the pope and his successors, so that he might be able to hear and be obedient to Christ. Yet Francis's own humility called him to an obedience that went beyond church authority. He wrote:

> Holy Obedience destroys
> every wish of the body and the flesh
> and binds its mortified body to obedience of
>   the Spirit
> and to obedience of one's brother
> and the possessor is subject and submissive
> to all persons in the world
> and not to man only
> but even to all beasts and wild animals
> so that they may do whatever they want with
>   him
> inasmuch as it has been given to them from
>   above by the Lord.[11]

Francis wanted to give himself fully to his relationship with creation, and thus he sought to be submissive to it and to encourage his followers to be submissive to it. In the late twentieth century, as the rate of species extinction surpasses one per minute, Francis's exhortation to make ourselves subject and submissive to beasts and wild animals takes on new urgency. The overarching reason for species extinction today is habitat destruction. If Franciscans are to concern ourselves with the survival of the diversity of the world's species, we must begin to defend the integrity and intrinsic right of "all beasts and wild animals" to have habitat, to have a home. To be faithful to our founder, Franciscans today must give voice to the cry of all creatures threatened with extinction and stand against the economic and political forces which encourage the destruction of their habitat.

The Canticle

Toward the end of his life Francis suffered greatly from ill health (he had gradually become blind, and some historians believe he suffered from tuberculosis) and also, some believe, from watching his fraternity grow into the thousands while losing some of its original passion for poverty and simplicity. During his last few years he spent over half his time in the wilderness, deepening his experience of contemplative prayer. He withdrew with a few companions from the complexities of his order's administration to seek God in nature, and it was there that he had his most profound mystical experiences, most notably the stigmata. The hermitages in which he dwelled were really no more than small caves or huts constructed of sticks, leaves, and mud, but they were located in areas of great natural beauty.

During this twilight of his life on Earth, Francis wrote his greatest piece of poetry, "The Canticle of the Creatures." During the spring of 1225, a little over a year before his death, Francis had a dream in which a voice spoke to him and encouraged him to praise God even in the midst of his infirmities. The Legend of Perugia tells us Francis arose the next day and said:

> I wish to compose a new "Praises of the Lord," for his creatures. These creatures minister to our needs every day; without them we could not live; and through them the human race greatly offends the Creator. Every day we fail to appreciate so great a blessing by not praising as we should the Creator and dispenser of all these gifts.[12]

Francis composed "The Canticle of the Creatures" in vernacular Italian during the period when this language was branching out from Latin, and this is the earliest recorded poem in Italian. For this reason, "The Canticle" has received a great deal of attention from philologists, with hundreds of articles treating it in this century.[13] Sadly, it has yet to be interpreted adequately by an environmental ethicist.

This poem reveals the summit of the spiritual journey of Francis as a Christian and a lover of nature. He communicates his vision of the interconnectedness of life and an inspiring mystical unity, all in a poem that is remarkable for its spontaneity and freshness. It speaks of the essential values of what is today loosely referred to as Franciscan Spirituality: the intrinsic goodness of all the created world, the interdependence of all life, a passion for beauty and for peace, and the personalism of Francis.

As Francis sang of the diverse parts of nature, he described them in remarkably intimate terms; he expresses great joy at having lived in relationship to them. He wrote:

> Praised be You, my Lord, by Brother Fire,
> through whom you light the night,
> and he is beautiful, and playful and robust
>   and strong.
> Praised be You, my Lord, for our Sister
>   Mother Earth,
> who sustains and governs us,

and who produces varied fruits with colored
flowers and herbs.

Francis saw the diversity of nature, and he sang its praises, seeing this diversity as a value in itself. In this century ecologists have begun to describe the previously unknown biological diversity of life and to warn us of the severe environmental consequences of not protecting it. We Catholics have the responsibility to preach the value of biodiversity of which Francis sang.

In his description of nature Francis used intimate, familial terms. He sang their praises because God was glorified through them, by them, and for them. He loved them because they complemented each other, making an orchestra of praise to our Creator. All forms of life, in all their marvelous diversity, are valuable and have a role in God's world. "The Canticle" is remarkable for the way it points out a principle ecologists have only recently begun to prove scientifically: all of life is interconnected. It is also suggestive of Paul's metaphor for the interdependence of the diverse members of the body in 1 Corinthians 12.

The word *for* in the above verse is translated from the Italian *per*, which can be translated "for," "by," or "through." All are acceptable, and the variety of meanings lends greater charm to the poem. I believe Francis was comfortable praising God *for* creation, praising God with and *through* creation, and witnessing God being praised *by* creation.

But Francis was also aware of the limitations of this life. The second stanza was written when the mayor and the bishop of Assisi were in conflict with one another; he composed it and had it sung for them so that they might achieve reconciliation. He wrote the last stanza on his deathbed, and it is believed to be his final work.

Francis and Deep Ecology: Where Do They Part Ways?

As shown above, there is much in common between the ecological philosophy of deep ecology and St. Francis. Both value the diversity of life and its intrinsic value. And Francis believed strongly in the gospel call to conversion, simplicity, and penance, values that are consistent with the third principle of deep ecology, that of achieving balance through simplifying and reforming our lives to emphasize the nonmaterial value of life.

In one respect Francis asks and responds to an even deeper set of questions, which Naess does not address: Why do human beings insist on destroying their habitat? What is it in our race that leads us to destroy other forms of life? In the context of Medieval

Europe, which understood Adam's Fall and original sin primarily in terms of disobedience, Francis understood that greed also played a part:

> For the person eats of the tree of the knowledge of good who appropriates to himself his own will and thus exalts himself over the good things which the Lord says and does in him, and thus, through the suggestion of the devil and the transgression of the command, what he eats becomes for him the fruit of the knowledge of evil.[14]

It is humankind's sinful nature, our grasping, greed, and disobedience, which leads us to break shalom, the peace of God. Only by repentance, or turning away from our sinful ways, can we find healing for ourselves and our relationship with creation and our Creator.

Most of the specific tenets elaborated by deep ecology are consistent with Francis's views as well. Wilderness has intrinsic worth, because without undisturbed wild places the diversity of life cannot be preserved. The love of a sense of place was a value for Francis, and he had great affection for several of his wilderness hermitages. Spirituality and self-realization were also very much a part of his religious journey.

Another principle of deep ecology is its opposition to industrial society, but it is more problematic to evaluate Francis's attitude toward this. While industrial society has helped many in the Northern Hemisphere attain a more comfortable standard of living, its consequences include the further impoverishment of the underdeveloped Southern Hemisphere. He certainly would be opposed to the large-scale environmental destruction being wreaked on nature in the form of the loss of biodiversity, ozone destruction, and industrial-scale deforestation.

Francis lived as capitalism was just beginning to take hold of Medieval Europe, and regional economies were rapidly converting from barter methods to currency. This enabled the accumulation of previously unknown quantities of wealth and power, which the Catholic church had traditionally opposed (with its stance against usury). Francis forbade his followers to possess coins, not because he wanted to deny them their basic needs, but because during the early thirteenth century coins were roughly equivalent to owning stocks and bonds today, and this would contradict his love of poverty.

Because he forbade his brothers to participate in the nascent capital economy of his time, I believe Francis opposed the concentration of wealth and power upon which industrialized society is

based. He understood that it would create barriers to human relationships, further marginalize the poor, and instigate threats to
peace. For this reason I conclude that he would agree with deep
ecology's notion that appropriate-scale, simpler lifestyles and economies are essential for a more humane society, one that fosters
spiritual virtues. Francis, like deep ecologists, looked to an alternative philosophy of life, differing from that of his dominant culture.

The thinking of Francis and Naess conflict most clearly over the
second basic principle of deep ecology: humans at present are far
too numerous and intrusive with respect to other life forms and
the living earth, with disastrous consequences for all, and must
achieve a substantial decrease in population to permit the flourishing of human and nonhuman life. Francis, of course, was neither
an ecologist nor a demographer, so even had he anticipated the
impending environmental catastrophes of this century, he could
not have analyzed their causes. First, however, I feel a quote from
the U.S. bishops in their pastoral letter on the environment can
help frame this issue.

> In public discussions, two areas are particularly cited as re-
> quiring greater care and judgement on the part of human
> beings. The first is consumption of resources. The second is
> growth in world population. Regrettably, advantaged groups
> seem often more intent on curbing Third World births than on
> restraining the even more voracious consumerism of the de-
> veloped world.
>
> We believe this compounds injustice and increases disre-
> spect for the life of the weakest among us. For example, it is
> not so much population growth, but the desperate efforts of
> debtor countries to pay their foreign debt by exporting prod-
> ucts to affluent industrial countries that drives poor peasants
> off their land and up eroding hillsides, where in the effort to
> survive, they also destroy the environment. Consumption in
> the developed nations remains the single greatest source of
> global environmental destruction.[15]

While the bishops do not adequately address the problem of overpopulation or prescribe any solution to the problems it presents,
they are correct in pointing out that there are basically enough
resources, at present, to meet the needs of everyone on the planet.
More important, however, they address the most grievous sins (from
an environmental standpoint) of the industrialized North: overconsumption and idolatrous greed. The North has no grounds to
criticize the South until it "removes the log from its own eye" (Mt
7:3-5) and repents of its ecologically ruinous practices driven by

industrial capitalism. Still, the time has long passed when the church should have rethought its failure to address the problem of population. We don't know what Francis would think about the population explosion. But we do know, based on Francis's unequivocal commitment to living and preaching peace in the context of a church bent on crusading, that he was willing to adopt unpopular positions. I suspect that Francis would agree with Naess that we humans are too intrusive with respect to other life forms on earth, but I seriously doubt he would feel that there are too many human beings. I suggest that, for better or worse, in his humility Francis would not presume to criticize the population growth of other cultures.

A more fundamental critique of deep ecology, however, would address its relationship to ideology. Naess supports "significant changes in ideological structures." If Francis were alive today, I believe he would say that it is precisely the nature of ideologies themselves that help create social and environmental problems; indeed, they help foment societal sin and sinful structures. Consider how the ideology of nationalism has helped bring about a society which is predisposed toward war and violence. The solution to problems created by selfish and destructive ideologies is not the restructuring of ideologies or a misanthropic ideology. The most compelling critique of ideology by Francis is demonstrated by his value structure, which moved beyond ideology.

St. Francis is famous for the value he placed on relationship and a personal touch. Much of his enduring charm can be traced to this. G. K. Chesterton, who wrote a biography of St. Francis almost forty years ago,[16] notes that Francis deliberately did not see the forest for the trees. In other words, he never lost track of an individual in a group, and for this reason Francis has been called the supreme personalist. He even personalized his relationship with Christ's poverty, referring to it as Lady Poverty, and exhorting his followers to be her knights errant. In "The Canticle" he writes about the elements as friends and family members, expressing his mystical vision of life's interdependence. He refers to them with terms of familiarity and affection.

This view can be contrasted clearly with the misanthropic attitude of some deep ecologists toward overpopulation problems. Dave Foreman, one of the founders of the Earth First! Movement, when interviewed about starvation in Ethiopia, said: "The best thing would be to just let nature seek its own balance—to let the people there just starve there." A different author wrote in the *Earth First! Journal* that, despite the suffering involved, AIDS was a "welcome development" in the necessary reduction of human overpopulation, especially since, unlike war and environmental catastrophe,

AIDS only affects human life. Even though these are careless statements publicized outside of their contexts, and even though the
Earth First! Movement has contributed much to the environmental movement, I believe such comments reveal a flaw in the thinking
and strategy of some deep ecologists. More concerned with ideology than with helping their fellow humans appreciate the gift God
bestows upon us in creation, deep ecologists settle (in frustration,
I believe) for misanthropy. Sadly, creation itself is betrayed by this
kind of hateful language, because people turn against the environmental movement as a result. A misanthropic ideology is no solution
to the problems posed by an anthropocentric ideology. Only the
transformative power of love and spiritual conversion can help us
recover a healthy relationship with nature. Anyone who truly cares
about the survival of nature's diversity on this planet must realize
that ideology and contempt for the human race will only alienate
people from our interest in nurturing a love for creation. Our path
must lead us to personal and social conversion, not a rearranged
ideology.

Recovering the intimate, sensual, personal relationship of Francis
with creation is essential if we are to arrest environmental destruction. Nature is already suffering from overpopulation and
overconsumption, and this will only get worse in the near future,
but I believe no rearrangement in ideology can initiate the improvements needed. What does have the power to transform human
conduct? It must be love, love for our fellow humans and love for
the fellow creatures on spaceship earth. Love had the ability virtually to eliminate slavery on a global scale in the period of a few
decades, and it had the ability to initiate dramatic changes in the
thinking of the United States regarding racism (although both required painstaking generations of effort before they were achieved).
In its effectiveness as a tool of persuasion, love far outshines ideology, and for this reason, if for no other, I believe Christians must
recover the relationship Francis had with creation. It has been given
to us as a gift, and it is now our responsibility to live it out.

Francis and Benedict: Brother and Steward

I believe deep ecologists are correct in their criticism of the stewardship model of understanding the human relation with the rest
of creation, at least as it is popularly understood in the United
States. Stewardship used to mean caring for things on behalf of
others, in this case, both our Creator and future generations, but
in the context of our North American culture's narcissism, we have
twisted the term to serve our own greedy desires. In extreme cases
the term stewardship has been corrupted by groups such as the

so-called Wise Use Movement to mean further ecological despoliation by unneeded resource extraction. I believe we must reform
our understanding of stewardship, fully aware of our society's predisposition toward myopic selfishness. I ask, indeed beg, my brother
and sister Benedictines to promote a stewardship that includes
selflessness and altruism. I pray that they can help our church
rediscover the prayerful and God-dependent roots of this model.
Stewardship should incorporate the understanding that nature has
a right to be cared for and safeguarded, not simply manipulated
for our desires.

But Francis offers us another way to relate to nature. As shown
above, he related to nature with altruism and humility, and he
respected and praised the way we are interdependent with creation. Francis valued nature for its intrinsic goodness, seeking to
submit himself to it, and I believe that if the integrity of our environment is to survive we must rediscover these ideals. All of creation
has the right to exist so that it can give praise to God; it need not
meet human needs to justify its continued existence. I refer to this
as the fraternal model of relating to creation.

The fraternal model holds that the Lord God has always been
involved in every aspect of the creation of nature and that as humans our role is to strive to live in harmony with creation by giving
praise to God. Presenting an alternative to the utilitarianism of the
stewardship model, the fraternal model emphasizes fraternity and
humility, the most important values Francis brought to his relationship with nature. Respect for the right of other species to live
in peace will propel adherents of this model to challenge their fellow humans to live within limits so that other species may simply
live.

The stewardship model has dominated Christian thinking because it explained theologically the way humans made contact with
nature, through agriculture. In the past two centuries the effect of
the Industrial Revolution has been twofold: it has brought a previously unthinkable quantity of land under human domination (with
the resulting ecological damage) and simultaneously resulted in
massive worldwide human migrations from rural to urban areas,
further removing us from nature. In the context of this changed
relationship, with a much greater proportion of human population
living in urban areas and a lesser proportion directly dependent
upon nature for living, the fraternal model can move us toward
greater friendship with the earth and more habitat protection. The
Catholic church must begin to respond to the massive wave of
species extinction now occurring. The fraternal model cannot replace the stewardship model, but as a complement it can help us
as Catholics draw from our own tradition to meet the challenges
posed by alienation from nature and guide our church's thinking

toward a sustainable future. We Franciscans have the obligation of promoting this model.

Wolf of Gubbio

The Little Flowers of St. Francis relates the story of when Francis visited the town of Gubbio while it was being terrorized by a huge, ravenous wolf.[17] Francis, filled with faith in God, journeyed out of the town, found the wolf, and began to preach peace to it. The saint commanded it to stop terrorizing the citizens, pledging that they would feed him and care for his needs. He then led the wolf into the town, preached penance and peace to the citizens, and forged a covenant between the humans and the wolf, bringing about happy reconciliation and peaceful lives. It would be impossible to verify whether these events ever occurred or whether Francis actually was able to work such a miracle, but my point in introducing this story has little to do with establishing either. I am more interested in the impact Francis had on his biographers; he was understood to be a man of peace and a preacher of reconciliation. He wanted the residents of Gubbio to be free from their fear of the wolf, so he exhorted them to reform their lives and live out gospel values. Francis did not seek to destroy or eradicate the wolf; he sought to become its brother and bring it into peaceful relationship with its environment. For our peace and well being, we humans must undergo conversion and strive for reconciliation with God, each other, and nature. This is the kernel of truth communicated by this story of the wolf, and to be faithful to Francis, we must seek to live out this message of reconciliation.

Francis and deep ecology communicate many compatible truths. The essential difference, it seems to me, is that deep ecology seeks to change ideologies, while Francis prefers a spirit of fraternity and respect. I hope that the twentieth-century followers of Francis can imitate this, freeing ourselves from anger and rancor, and freeing ourselves for love, love for the soil, the flowers, the micro-organisms, the birds, the wild animals, for the diverse, incredible, intricate interconnectedness of life that St. Francis sang about with such joy. St. Francis, pray for us!

The Canticle of the Creatures

I.
Most High, all powerful, good Lord,
Yours are the praises, the glory, the honor,

and all blessing.
To You alone, Most High, do they belong,
and no one is worth to mention Your name.
Praised be You, My Lord, with all Your
        creatures,
especially Sir Brother Sun,
Who is the day and through whom You give
        us light.
And he is beautiful and radiant with great
        splendor;
and bears a likeness of You, Most High One.
Praise be You, my Lord, by Sister Moon and
        the stars,
in heaven You formed them clear and
        precious and beautiful.
Praised be You, my Lord, by Brother Wind,
and through the air, cloudy and serene, and
        every kind of weather
through which You give sustenance to Your
        creatures.
Praised be You, my Lord, by Sister Water,
which is very useful and humble and
        precious and chaste.
Praised be You, my Lord, by Brother Fire,
through whom You light the night,
and he is beautiful and playful and robust
        and strong.
Praised be You, my Lord, by our Sister
        Mother Earth,
who sustains and governs us,
and who produces varied fruits with colored
        flowers and herbs.

II.

Praised be You, my Lord, through those who
        give pardon for Your love
and bear infirmity and tribulation.
Blessed are those who endure in peace
for by You, Most High, they shall be crowned.

III.

Praised be You, my Lord, through our Sister
        Bodily Death,
from whom no living man can escape.
Woe to those who die in mortal sin.

Blessed are those whom death will find in
    Your most holy will, for the second death
    shall do them no harm.
Praise and bless my Lord and give Him
    thanks
and serve Him with great humility.

Notes

1. Lynn White, Jr., "The Historic Roots of Our Ecological Crisis," *Science* (March 10, 1967), p. 1203.

2. Naess, Arne, "Intuition, Intrinsic Value, and Deep Ecology," *The Ecologist* (England), vol 14, no. 5-6 (September/October 1984). The most comprehensive work on deep ecology is Devall, Bill, and George Sessions, *Deep Ecology* (Layton, Utah: Gibbs Smith, 1985).

3. The best treatment of this problem is found in Roger Sorrell, *St. Francis of Assisi and Nature* (New York: Oxford University Press, 1988).

4. Kucak, Tanya, "Deep Ecology," *Yoga Journal* (September 1986), p. 36.

5. Thomas of Celano, *First Life of St. Francis*, 6. All the biographies are taken from *St. Francis of Assisi, Writings and Early Biographies*, ed. Marion A. Habig O.F.M. (Chicago: Franciscan Herald Press, 1972), which is commonly referred to as the Omnibus of Sources. All numerical references to these biographies of Francis refer to paragraphs, not pages. Celano was a friar, a contemporary of Francis, and an eyewitness to some of the events later in his life. His was the first biography of Francis.

6. Thomas of Celano, *The Second Biography of St. Francis*, 165.

7. Ibid.

8. Thomas of Celano, *First Life of St. Francis*, 80.

9. Ibid.

10. Francis of Assisi, *Rule of 1223*. All of Francis's own writings are taken from Regis J. Armstrong, O.F.M.Cap., and Ignatius Brady, O.F.M., *Francis and Clare: The Complete Works* (New York: Paulist Press, 1982).

11. Francis of Assisi, "Salutation of the Virtues," verses 14-18, in Armstrong and Brady.

12. Legend of Perugia, 43, in the Omnibus.

13. See Armstrong and Brady, p. 37.

14. Francis of Assisi, "Admonitions," para. 2.

15. United States Catholic Conference, *Renewing the Earth* (Washington, D.C.: U.S.C.C, 1991), p. 9.

16. G. K. Chesterton, *St. Francis of Assisi* (Garden City, N.Y.: Doubleday, Image Books, 1957).

17. *The Little Flowers of St. Francis*, 21, in the Omnibus. This is a later biography, composed at least one hundred years after the death of Francis. Most scholars question its historical accuracy.

18

Choose Life

Ascetic Theology, History, and Ecology

DAVID M. SHERMAN

ALSO KNOWN AS H.H. BHAKTI ANANDA GOSWAMI

I have set before you life and death. . . . Choose life that you and your descendants may live. (Deuteronomy 30:19)

Without God, the Human Race Is a Threatened Species

What has been the historical role of Catholic asceticism in relationship to ecology and the spiritual, moral, and physical health of individuals and society? To answer this question we must confront the tragedy that much of Christianity has widely deviated from the ascetic moral discipline of the Apostolic Fathers. This failure in moral leadership has largely contributed to the fact that the human race is a threatened species, and that it is the pollution of our sin that is destroying us. The ancient Hebrew and Sanskrit word for defilement due to sin is *tame/tama*, to be "polluted" or to be "in darkness," "defiled," or "ignorant." In Old Testament times the righteous one (*tsaddic, sadhu, sattvika,* or "ascetic") was to battle sin, the cause of all filth/pollution. The English word *ascetic* derives from an ancient Greek word used to describe persons who practiced a religiously motivated discipline (called *sadhana* in Sanskrit). In ancient times asceticism was originally associated with the widespread, multi-ethnic religious alliances of the Old Testament Eli-Jahu (Vaishnava Hari-Vasu).

With the coming of the gospel, Jesus Christ was received as the promised redeemer of Jew and Gentile alike. Within two hundred years of the age of the apostles, the Mediterranean regions' Sacred

Mystery Feasts of Eli-Jahu's Messiah transformed into the Mass of Christianity. The revelation of God's saving grace in Jesus Christ was accepted as the completion and perfection of previous revelation. The previous Eli-Jahu alliances were replaced by Catholicism.

The doctrines, ascetic morality, sacraments, and social structure of the ancient Eli-Jahu alliances continued in the early church. Today these same elements of the ancient Eli-Jahu religion are found preserved to an extraordinary degree in the related major Eastern religions of Vaishnavism, Pure Land Buddhism, Sattvic Theistic Shaivism, and in smaller religions derived from these. Scholars have long noted profound similarities to Catholicism in these traditions, but no explanation has been acceptable to all parties. From the interdisciplinary evidence I have amassed over thirty years, it is clear that these traditions are closely related to Catholicism by descent from the same ancient *sampradaya* or teaching lineage of Messianic Eli-Jahu worship. Presently these related traditions represent approximately a billion souls prepared by Eli-Jahu to receive the gospel of Jesus Christ. However, the deviation of Christian culture from apostolic asceticism poses an enormous barrier to the spread of the gospel among these peoples, whose moral ideal and general lifestyle remain ascetic.

During intertestamental times, asceticism continued among Eli-Jahu's faithful, but it was also found in the Mediterranean and Asia among many adversarial groups, including anti-Yahwist gnostics and atheists. Asceticism alone does not guarantee love of God and neighbor. Today we have much detailed information on the various ascetical moral disciplines and the different groups that practiced them. The Nazarite Jews and first Christians were ascetic. Catholic asceticism is a continuation of this particular spiritually motivated practice of both mental and physical moral discipline. Some of this discipline focuses on the control of appetites and consumption in relationship to the Christian stewardship of our bodies and the environment.

Historically the asceticism of the aforementioned related Eastern Eli-Jahu believers has had an enormous impact on their relationships to each other, to creatures, to their economic activity, and to the environment. Today, much of their individual, family, and community life focuses on devotional service, works of redemption, restoration, and mercy. Many embrace Jesus as a *sant* (saint), *purohita* (prophet-priest), or *sadhu* (Jewish *tzaddik*), and some even adore Him as God; however Christianity is not respected as having moral authority because it is perceived as anti-ascetical, exploitative, and sinfully indulgent.

What has this to do with the fate of the earth? What is wrong with the value system that has brought us to the present ecological impasse? We cannot separate the morality of humanity from the vitality of the biosphere. The more we sin, the more earth is cursed due to that sin. Surrendering to the Creator's good moral order restores us to a state of grace, healing our relationship to God, one another, and to the whole community of creation. A revival of inspired asceticism is needed to renew the moral leadership and spiritual vigor of Christianity and to open the door to unity with our Eastern relatives. This spiritual renewal and unity is the key to global cooperation in restoring the vitality of the biosphere.

"Take not thy Holy Spirit from me" (Ps 51:13)

The good order of Eli-Jahu's law (*Torah/Dharma*) is inspired and sustained by His Holy Spirit. The Holy Spirit has spoken through the prophets and prophetesses. Inspired souls reveal God's law, rebuke evildoers, warn the faithful, announce the coming of the Messiah, or perform some other sacred service, preaching, or priestly function. These souls are *Dharma* protectors, who lead God's people in their battles against the evil order of the enslaver antichrist. There are two enemies of good order, chaos and evil order. Evil order is organized crime, the devil's functional facsimile of God's good social order. The devil cannot create, so he has to be satisfied with hijacking parts of God's good creation and retooling them for his own cruel use. Prophets expose these intrigues of the devil. Today evil is overtaking humanity, and yet God's people fail to convict and halt the sinful activities of the perpetrators. Purity is potency. Spiritual impotence is due to sin.

The guilty do business openly in the international marketplace selling everything from legal intoxicants to ill-gotten human organs and "harvested fetal tissue." Legal abortion has murdered approximately 100 million infants since the 1960s. Millions of female genital mutilations have been legally performed in African and Muslim nations. Legal arms trade has turned the earth into a high-tech battlefield. There are presently 100 million land mines deployed on the earth. When legal commerce is so horrifically perverse, what must be the depth of evil, the social and environmental cost of illegal activity? People of goodwill and the earth itself are thus victimized by powers of darkness in high places, trapped between these social predators and parasites as they battle with each other over territorial disputes.

Hunger and the Environment, the Evils
of Predatory Agribusiness

In ways hidden to most who live in meat-eating societies, modern agribusiness diverts grain (the staff of life) from human mouths
on a scale that defies the imagination. Appalling statistics on the
social and environmental costs of livestock agribusiness can be
found in the reports of the U.S. Department of Agriculture and
detailed in books like those of Frances Moore Lappe (*Diet For A
Small Planet*) and Jeremy Rifkin (*Beyond Beef*). Considering that
the statistics on American livestock grain waste must be multi-
plied to assess the global cost of such practices, the implications
for the human hunger dilemma stagger the mind. Over 120 million
metric tons of grain are completely wasted in American livestock
production each year. Two-thirds of the grain exported from the
United States goes to feed stock. Millions of acres of third-world
land are used to produce feed for European livestock. This squan-
dering of agricultural resources is a crime against humanity, nature,
and God.

The ecological cost of such livestock production is incalculable.
Let us recall a basic moral teaching rooted in the ascetical tradition of Catholicism. In that catechesis we learn that we share guilt
in a sin if we stand by and acquiesce in it. On this principle of guilt
by being accessory to another's sin, I cannot take seriously the
commitment of any environmentalist who, knowing the social and
environmental costs of livestock agribusiness, still continues to
participate in this sin by choice. There are presently over one billion bred-for-slaughter animals on the earth. These are consuming
vast amounts of food and fresh water, most of which is completely
wasted in the metabolic process. Unless gluttonous humanity faces
this issue of unnecessary animal breeding and sacrifice to the false
gods of acquired taste, it shall literally eat the biosphere to death.

We worry about deforestation, pollution, and about fresh water
supplies. Fossil-fuel burning has our attention, but if we go back
to sources like Rifkin's book we see that nearly half the water used
in the United States grows feed for cattle and other livestock. Lappe
has said that the water used to produce ten pounds of steak could
supply a family for a year. Annually, cattle in the United States
produce nearly one billion tons of waste and total livestock waste
accounts for twice the amount of pollutants as originate in all United
States industrial sources. A standard 10,000 head feedlot produces
nearly 500,000 pounds of manure in twenty-four hours. This equals
the human waste equivalent of 110,000 people. Much of this ends
up as runoff, polluting fresh water sources with dangerous nitrates
and other chemicals. Feedlots are the source of more than half the

toxic organic pollutants found in fresh water (statistics above from Rifkin). Add to this the amount of metabolic heat and greenhouse gases produced by livestock, and the global potential for environmental disaster becomes apparent.

If we love our planet, care at all about the availability of grain for human consumption, and have any sensitivity at all for living things, how can we let this wasteful orgy of gluttony go on? Have we Westerners been mesmerized by the romantic myth of the cowboy for so long that we cannot see the writing on the wall? Is it too late? Has modern global agribusiness become an unstoppable pusher of flesh, alcohol, tobacco, and other harmful addictive substances, pandering to the unrestrained desires and tastes of the world's wealthiest consumers? The economic pusher-suppliers and their addicted customer-slaves are committing crimes of consumption against humanity and the biosphere. In their godless, predatory global economic system, mundane wealth is power, and power controls the distribution of the earth's resources. Consequently, the powerless go without while nonrenewable resources are plundered sinfully to satisfy the self-destructive cravings of a conscienceless consumer elite. The have-nots die without, while the gluttonous die of the diseases of excessive and perverse consumption.

On a much smaller scale, this has been going on ever since Eden, when the natural order was first disrupted by humanity's illicit craving. Since then, the lust and gluttony of the few have often disrupted and deprived the many. On the large scale, agrarian civilizations have been repeatedly destroyed by predacious barbarians. Range wars between herd hunters (cattlemen) and settled farmers (including dairy farmers) have occurred often in history. The underlying cause is a clash of two opposing lifestyles, one that is earth-nurturing and one that is exploitative and predacious. Nurturing farmers belong in the Garden of Eden. They want lives of gentle peace, of daily and yearly cycles unbroken by crisis or the aggression of creatures large or small. They want to stay in one place and care for the land, which must remain fertile or they will die. They guard the source of water for their families to keep it pure and abundant. Gardeners are students of nature and careful stewards of resources; they aim at a sustainable life in a fixed place. Friends of the earth, they want generations of their families to enjoy even more abundance and variety from the land which has become Eden-like in their laborious care. Their efforts are in a way restorative penance for the pollution of Eden. Gardeners love the land they tend with their sweat, blood, and tears. It is a sacred trust, a second chance given by their Creator.

The perennial enemy of the nurturing gardener is the predatory, plundering herdchaser. Such conscienceless barbarians pulverize everything in their path as they drive their prey before them. Like the destroyers of ancient agrarian cultures, like lions or wolves chasing a herd, today's conscienceless factory farmers and cattlemen have no concern beyond the availability of their victims and the profit they will receive at market. When one verdant field is feed-farmed or grazed out, when one pond or aquifer is ruined, they just move on to the next. Denied cheap grazing on U.S. Federal land, they level a virgin rainforest or dispossess a poor people somewhere and soon are back in business. They take what they want from the environment and give nothing (by comparison) back. Small-scale subsistence hunting and fishing by indigenous peoples is *not* a threat to global ecology. It is the institutionalized evil of predatory agribusiness that is destroying the fertility and biodiversity of our planet. Power and technology wedded to gluttonous addiction and commerce without morality have produced this biological atom bomb of voracious livestock. Human lust and greed have created a population explosion of domesticated animals that is devouring and fouling the biosphere.

There are not too many human beings, only too many eating flesh at the top of the food chain instead of plants on the bottom of the food chain, which they were created to eat. This is exhausting the fertility of the earth. The problem is that we cannot just move on to the next planet when this one is feed-farmed and grazed out. We cannot follow our herds and flocks to a new extraterrestrial source of fresh water. We cannot feed all this unnecessary livestock and humanity too. The situation is a socio-economic, health, and ecological nightmare.

The Ascetic Angle of Vision on Our Predicament

The social and environmental disorder that human addiction to overconsumption causes can be studied the world over and throughout history. The ascetics and holy ones of many lands have understood the relationship between diet, personal, social, and environmental problems. Wise ones have observed dietary cause and effect played out in their own cultural milieu. The effect of diet on moral character, spiritual life, and society has always been an important issue in asceticism, and it suffices to say that the coming ecological crisis will not be averted until humanity faces this problem of gluttonous and unnecessary flesh consumption and solves it.

Pathological consumption means more disease and waste. Now sickened consumer societies are drowning in offal and looking for ways to get away from their various forms of excreta. Thus it is that hundreds of millions of lives are diminished and destroyed by the sinfully consumptive behavior of an economically powerful but deluded and enslaved human minority, whose waste and by-products of a toxic lifestyle have defiled the whole nest of humanity.

Addiction and Denial

These consumer-addicts are not criminals of the same order as their pushers. Although their consumption is immoral, they suffer from a diminished capacity to comprehend this because they are in an intoxicated condition. Their pathological consumption usually begins as a venial sin, but once addicted the addiction takes over. As the saying goes: First the man takes a drink, then the drink takes a drink, then the drink takes the man. These addicts are sick and need detoxification. They are possessed and need deliverance. They are rebels who need repentance and sinners who need salvation. They are failures who need forgiveness and liars in denial who need to hear the Truth that can set them free. Addicts are slaves to their pushers. Living in denial, addicts imagine their pusher as a friend or figure of providence and protection. They cannot imagine living without their addiction, and so they go on allowing these evil-doing parasites to feed off of them. Addicts live on illusion and protect their slave masters.

To the addict, God's prophet boldly declares: "You are the pathetic addict-slave of a cruel master." But convicting and rebuking addicts will not solve the problem. We must offer them salvation. We must intercede to set them free! We must also go after the pushers themselves where they seem safe within their bureaucratic, occult, institutional, social, economic, and false religious structures. God (Jahu Sabaoth) goes before his fearless people in the destroying of evil order, tearing down edifices and strongholds, liberating slaves, ending human sacrifice, and cleansing and restoring the holy places of the Lord. God's people should not be the slaves of false gods or enslavers but liberators!

As Catholics, part of a living tradition of Judeo-Christian asceticism, we should not have participated in crimes of consumption against humanity, nature, and God. Two thousand years of a vital ascetic Christian witness in the world might have even prevented this crisis of our biosphere. As Catholics we realize that grace is unmerited, not won by works of righteousness, but morality is one appropriate response of thanksgiving to God's grace. God calls us to worship Him in the beauty of holiness. Morality is a prerequisite to holiness. Moral temperance requires that even our natural desire for good things must be controlled as to quantity, time, and circumstance. Gluttony is the sin of destructive consumption.

Through ascetic discipline, we avoid gluttony by bridling the mind and senses and exercising our ability to say no to them. Having attained mental and physical control, we are able to resist when pushers tempt us to taste forbidden fruits. Freedom depends on this self-control. Without it we become slaves to our bodily appetites and to those who supply our self-destructive addictions. This bondage can diminish our ability to respond to God's grace. Unless we cultivate the ability to say no to both the kind and quantity of what is available for our consumption, we live without discipline like two-legged wild animals, completely at the mercy of our base wants and needs. Being civilized, mature adults should mean that we are in control of our appetites.

In reality, many non-Christians are highly civilized, while the Christians seeking to evangelize them are not. This is scandalous. While Christians should be models of ascetic virtue and leaders in moral sensitivity and stewardship, large numbers of them are slaves to the false gods of the tempter's world. Some of these addicts have become pushers of consumerism themselves in order to support their own perversely pleasurable but destructive habits. How is it possible that the pious, gentle, ascetic church of the apostles has appeared to permutate into the voracious, violence-ridden, materialistic, so-called Christian culture of today? Where and how have we Catholics deviated from the ancient moral ethical standards of our faith? What has happened to our asceticism, our moral teaching, our discipline?

Our Ascetic Heritage

Asceticism was the common discipline that incarnated the moral theology of the pre-Advent, non-ethnocentric, salvific, incarnational monotheism devoted to Eli-Jahu. Eli-Jahu's ascetics lived lives of intercessory penance for the sake of others' salvation. They sought to participate in the redemption of humanity and the restoration of earth. Whether they were Greek ascetics, Jewish *tzaddics*, Vaishnava *saddhus*, *sadhakas*, *sattvatas*, or Pure Land Buddhist *bodhisattvas*, these *sants* (saints) took an active role in the divine effort to restore enslaved humans to their rightful dignity as children of God. Like these pre-Advent Eli-Jahu believers, later Christian ascetics abstained from unnecessary killing and the consumption of flesh sacrificed to false gods. They abstained from intoxication, gambling, and sexual activity other than lawful wedded union. These activities were seen as debasing and polluting to both the inner and outer world of humanity. In our time, the Vaishnava Mahatma Gandhi taught these principles as *ahimsa* (nonviolence), *dhira* (sobriety), no wealth without work, and *brahmacharya* (chastity). These are four cardinal principles of ascetic morality. Christian love finds expression in these four principles as love-mercy, sobriety-self-control, purity-chastity, and integrity-truthfulness. Asceticism helps humans attain to these Christian perfections through the principle that abstinence from certain activities allows the development of opposing virtues. Virtue makes possible the realization of personal and social good. Believers are to cultivate the higher taste for goodness, which leads to holiness in spiritual maturity.

God's love for creatures is experienced in the healthful goodness of His holy benevolent order. Thus, believers are called to rise above an animalistic level of existence. God loves us all just as we are, but base behavior can harden our hearts toward one another, creatures, and God. Moral discipline gentles and civilizes us, making us more able to respond to God Who Is Love in a reciprocating loving manner. Asceticism can prepare us to be filled by the Holy Spirit, responding optimally to God's grace.

For the virtue of mercy to be cultivated, unnecessary killing and flesh eating should be avoided. Sobriety and self-control cannot be attained until gross and subtle forms of addictive intoxication are overcome. Holy love and the virtues of innocence and purity are destroyed by illicit exploitative sexual activity. Good stewardship, truthfulness, and integrity are undermined by all forms of unnecessary risk-taking and gambling, futures speculation, usury, the desire for wealth without honest work, and so forth.

The Kingdom of God Within and Defilement Without?

Sins of violence, intoxication, exploitative sexual activity, and gambling mark a society lost in the dark ignorance of anti-asceticism. Societies that revel in such internally self-polluting activities typically care little about polluting the outer world around them. How can those who carelessly poison their own bodies be expected to care about poisoning somebody else?

For enlightening discourse on the nature of societies in the polluted "mode of ignorance" (*tamo guna*) I recommend studying Vaishnava scriptures like the *Bhagavad Gita*. By comparison, the Gita describes societies in the "mode of goodness" (*sattva guna*) as characterized by compassion and nonviolence, control and orderliness, holy love and purity, truthfulness and integrity. Clearly Eli-Jahu's people are called to the ascetic morality of *sattva guna*. Jesus Christ still desires His beloved to clothe herself in the beauty of holiness and arise from the animal level of existence to claim the goodness of His promise. Jesus wants us to be virtuous not voracious, peaceful not violent, humble not arrogant, and spiritual not materialistic, racist, sexist, and so forth.

The Hope of the Faithful

First-century Christians sought to live again the pristine morality of the Edenic life before the Fall. They looked forward to Jesus Christ's return and the fulfillment of the Messianic prophecy in Isaiah, chapter 11:

Peace in Christ's Kingdom

[6] The wolf shall dwell with the lamb: and the leopard shall lie down with the kid: the calf and lion, and the sheep shall abide together, and a little child shall lead them. [7] The calf and the bear shall feed: their young ones shall rest together: and the lion shall eat straw like an ox. [8] And the sucking child shall play on the hole of the asp: and the weaned child shall thrust his hand into the den of the basilisk. [9] They shall not hurt, nor shall they kill in all my holy mountain, for the earth is filled with the knowledge of the Lord, as the covering waters of the sea.

The faithful often lived like the Nazarite ascetics under temporary or permanent vows and practiced temperance, abstinence, fasting, simplicity (voluntary poverty), purity, modesty, truthfulness, and renunciation of many "worldly" pleasures. Other forms of restraint like nonviolence and silence were also practiced. Constant prayer, humility, charity, a positive service attitude, and other virtues were cultivated in the early church. What has happened to this asceticism over the last twenty centuries? Is there still hope for humanity's unity and perfection in Christ? Are we still seeking the promise of God's benevolent kingdom?

Pathological Inculturation, Trials, and Tribulation

Somehow, in the rush to save souls the moral discipline of the early church broke down. The result was a separate set of behavioral expectations for the merely baptized or confirmed faithful and those who took additional vows of religious life. Often adjusted expectations resulted from Christian confrontation with uncivilized cultures. The development of certain regional strategies to cope with particular social problems resulted in a pathological pattern of moral compromise and inculturation that did not ultimately serve the global cause of Christ very well. Earth's present moral-ecological crisis is partly the poisonous fruit of this pathological inculturation. While some evangelized peoples continued in their newly refined, church-polished, covered barbarism, Christians in holy orders (priests, monks, and nuns) still taught and exercised spiritually via the ancient disciplines. As time passed, the ratio of religious to laity changed and the moral gap between them widened. Eventually, for many the idea of salvation was more connected to the formal profession of doctrine than it was to the subsequent practice of living Christ's morality in a continuing state of His grace.

An example of this change-process in disciplinary expectations can be seen in relationship to dietary temperance and the use of alcohol. First, dietary concessions were made for missionary religious living in the hostile environments of Northern Europe, where game was often the only food available in the midst of long, harsh winters. General flesh abstinence was not only impractical, but often impossible in some mission fields. Necessary adjustments in the ascetic rules were made. In some places, due to a lack of potable water, beer or wine was regularly mingled with it to kill germs. Such exceptional and justifiable modifications seem to have been the beginning of the end for the ancient apostolic practices of general abstinence from flesh foods and alcohol. Over the centuries, human frailty, time, and circumstance conspired to change slowly the ancient standard of abstinent and temperate behavior for believers. Nazarite Jews had abstained from consuming flesh and alcoholic wine. Knowledge of such abstinence, common among the Apostolic Fathers, finally disappeared from popular Western Christian memory. Then the slow cultural corruption of much religious life insured that it would remain forgotten for many in the church.

In time, a class of believers was created that was not expected to live up to old apostolic standards. Religious were still expected to, but eventually the standard itself was modified by religious for various justifiable and spurious reasons of survival and inculturation. These modifications then became a standard unto themselves and various rationales were created to defend the once exceptional practices as if they were actually part of our authentic Judeo-Christian moral heritage. Anti-ascetical Christianity is still suffering today from a pathological denial of this moral corruption. This nominal Christianity has become more and more locally confounded with the peculiar sins of cultures, races, and nationalities. This corrupt Christianity does not transcend such things as racism and sexism, but rather, it has been pressed into the service of these things. God's authority in Jesus Christ has now been invoked by evil-doers to justify their offenses against everyone and everything that they have sought to oppress and exploit. Christians, once known as pious ascetics in the age of the Apostolic Fathers, ended up running the breweries and wineries of Europe. Benedictine Rule accommodations for religious alcohol consumption and flesh-eating were stretched to the extreme. Ale became the main daily drink of the religious at monasteries like Durham in fifteenth-century England. Feast days found the tables of wealthy European religious spread with spiritous drinks and carcasses of every description. This was the beginning of the pathetic caricature of the gluttonous, drunken, carousing friar.

The Shameful Record of Our Sins

Because of pathological inculturation and general moral infidelity to Christ, for much of humanity, Western Christian culture has become identified with imperialism and economic predation, violence, enslavement of women and people of color, the genocide of non-Christian peoples, and the worldwide development of markets for opium, heroin, cocaine, alcohol, tobacco, coffee, and other addictive substances. Christians have sinned by tolerating these evils and promoting sexism, racism, fanatical nationalism, and other evils in the name of Christ. Must Christians today take up Oriental ascetical disciplines to learn how to control their minds and senses? Must we stand ashamed in the presence of pious, gentle "pagans"? Can't we see that a drinking, smoking, gluttonous, and sexually immoral priesthood will fail utterly to inspire faith in the "un-saved" and holiness in the faithful? Does "carne-val," the feast-of-the-flesh, really belong anywhere in the Christian life? No. We need to return to our own Judeo-Christian ascetical roots to recover the moral authority and spiritual power of God to overcome these evils in our individual lives and in our societies.

Asceticism and Catholic Charismatic Renewal

Ascetic theology commands us to "be filled with the Holy Spirit!" Authentic Catholic Charismatic renewal will be evidenced by the presence of all the gifts of the Holy Spirit filling the members of His church. As Abba Matthew the Poor states in *The Communion of Love* (1984):

We always proceed from theoretical statements in dogmatic theology to practical applications in ascetic theology when we come to deal with the Holy Spirit.

. . . and so if we are really listening to the Holy Spirit as our conscience, then

. . . every gift from God granted us freely by faith is transformed in us into commitment to perfect it in action.

Christian asceticism is the soul's striving to be filled with that perfection found in Christ Jesus, by the grace of His gifts in living a Holy Spirit-filled life. To be filled with God, we must first empty ourselves of all ungodliness. The spiritual discipline of asceticism is the housecleaning of our soul that prepares our heart for the indwelling of the fullness of Christ's perfection. All Christians are called to such perfection, not just "religious." The spiritual and moral discipline of asceticism is necessary to perfect the likeness of Jesus Christ in us and to release the healing, lifegiving gifts of His Holy Spirit into our sick and dying world. Asceticism is a continual striving, because the pushers of the world are always at our door trying their hardest by all means to invade God's rightful space in us and occupy it with their own sources of limited, temporary, and perverted satisfaction. Every day we must be prepared anew to drive these gangs of plunderers, tempters, and con artists from our individual and collective door. Asceticism is the exercise we get defending the borders of our soul and keeping our spiritual house in order as a clean and pleasant dwelling place for our God Who Is Love.

Ascetic discipline is not undertaken to earn salvation. Rather, Christians seek perfection only out of love, in order to become more pleasing to their Lord who loves them. There is no selfish or oppressive motive, and no mundane desire for power and control in pure mature Christian asceticism. Saints are not proud of their ascetic accomplishment. Love of God, and desire to conform to His goodness in all things is the motivation for everything committed in or omitted from a Christian ascetic's life.

Asceticism orders the internal life and external activity of persons according to the will of God, who exists as a Trinity of Divine Persons eternally related in holy love. Because there is this communion of love within the interiority of the Trinitarian Godhead, God's creative expression in the world has made us to be creatures of his loving community here as well. Ecology is concerned with the restoration of Edenic life-giving relationship within the earthly community of creation. The fundamental nature of sin is that it destroys right relationship in the community of God's love. The disorder of sin in the human microcosm expresses itself in the malevolent disorder of societies and the macro-disordering of the ecosystem. Personal sin, as offense to the Creator God Who Is Love, cannot be isolated from social and environmental consequences.

There can be no restoration of the human family or environment as long as responsible humanity remains in denial of its sin. Sin must be named and confessed to be exorcised. Repentance and ascetic penance is the divine prescription to restore the spiritual and physical health of humankind and the earth. The most appropriate acts of penance are those of reparation and restoration. Humanity must restore the earth. Divine justice demands it, and God's mercy in Christ will make it possible if we repent of our sinful ways and follow Him.

The Good Law, God's Prescription for Our Mortally Polluted World

Mahatma Gandhi summed up traditional Vaishnava social teachings in the following way. He gleaned seven root causes of social evil from his studies in scripture and in life: 1) Wealth without work; 2) Pleasure without conscience; 3) Knowledge without character; 4) Commerce without morality; 5) Science without humanity; 6) Worship without sacrifice; 7) Politics without principles. Gandhi had a great love of the Bible and the Vaishnava scripture, the *Bhagavad Gita*. As a Vaishnava, he recognized his own faith's moral precepts in the teachings of Jesus Christ. One of Gandhi's favorite passages in the *Bhagavad Gita* was Chapter 16, verses 1 to 3, in which God describes the qualities of a believer as fearlessness, purity, wisdom, unity/love, charity, self-control, self-sacrifice, austerity, penance, simplicity, nonviolence, truthfulness, freedom from anger, renunciation, tranquility, aversion to faultfinding, freedom from greed, gentleness, modesty, determination, vigor, forgiveness, fortitude, cleanliness, freedom from envy, and freedom from the desire for honor (*Bhagavad Gita As It Is*, B.B.T Edition, pages 721 and 722).

These are the universal attributes of Eli-Jahu's saints and the qualities we must all seek to attain if we want to be part of the solution instead of part of the problem today. Let us all pray for the courage to strive for this high calling in Christ Jesus. Amen.

19

Concluding Reflections

Toward a Second Axial Age

WAYNE TEASDALE

Each of the contributors to this unusual volume has a passionate commitment to the future of the earth, to environmental awareness, or to ecological justice. All of us have a deep appreciation of ecological spirituality or are able to discern the presence of the divine in, through, and enveloping the cosmos. We are all trying to spark this same passionate concern for our planet, the entire world, including the human species, in the church and in society. We want it to become a major focus of ecclesial leaders as it is becoming of political figures. The task for us all is—to inspire change in society, culture, our church, in how humans regard and relate to the environment or the earth. There is genuine concern for and interest in the issue of environmental preservation among Catholics, but it has not yet reached the level of passionate commitment, nor has awareness of the spiritual value of the earth dawned in our leaders.

An anecdote from my own experience sheds light on this problem. I met Thomas Berry for the first time in 1977 at his Riverdale Center in the North Bronx, New York. I was then a graduate student at Fordham University. I had then and still have now—hopefully in a more mature form—a profound interest in mysticism. My enthusiasm for mysticism was, indeed still is, as passionate as Thomas Berry's passion for the earth. I encountered Tom in the backyard of his center in front of an extraordinary archetypal tree that is hundreds of years old. I was trying to engage him on the topic of God and then mystical experience. Tom remarked, "Well, God is not a category I think in." I was stunned, since he is also a priest. So, I responded, "Oh, I wasn't aware that God is a category!"

My somewhat flippant response did not address Tom's sober urgency of focus, the context that inspired his statement on God. I went on to read many of his Riverdale Papers, and I recall how at the time I regarded his New Story as a likely story. I was so absorbed in the mystical that I couldn't grasp the seriousness of the issue other than in an abstract intellectual sense. Tom's point seemed theoretical to me. I failed to grasp its importance because I was asleep, secure in a kind of other-worldly spirituality. Over the years I have gradually adapted a more balanced approach, one that attempts to reconcile the two realms, realms that are actually more integrated than they may appear.

Each contributor to this book accepts either explicitly or by implication the extreme poignancy of Thomas Berry's question at the beginning of this work: "After we burn our lifeboat, how will we stay afloat?" Catholics have a profoundly beautiful tradition, a spirituality that is unassailable, and the precious treasure of the eucharist, but as Thomas Berry points out so effectively, these function within the reality of the earth itself. Religion is not, nor can it ever legitimately be, separate from the earth. Berry's question is a challenge to Catholicism, and he asks it because the Catholic church has been a major source of our failure to respond to the environmental crisis, along with the rest of Christianity, not to mention the other world religions. Berry's challenge is meant for religion as a dimension of human experience and more important, as an institution in the world that makes decisions that negatively or positively affect the environment. As his challenge is taken more to heart in our tradition, the difficult path to negotiate—difficult because so new—will be how to remember God while embracing the earth.

Most are convinced of the necessity for deep ecology, the subject of Keith Warner's challenging article, and of the powerful insight that the human is part of nature—as well as a pernicious obstacle to planetary healing. Albert LaChance's chapter builds on Mircea Eliade's understanding of the relationship of ancient cosmologies and cosmogonies and of the evolution and health of cultures. In the modern world cosmology as the study of the origin and nature of the universe has been cut off from cosmogony, the study of the divine role in creating the universe. Heretofore cosmology and cosmogony were in agreement, and all cultures developed out of this inner balance. Cosmogony revealed the sacrality of cosmology. Because this relationship has been ruptured, we are witnessing the disintegration of cultures. The final weakness of contemporary cosmology is seen in the way it leaves God and Christ out of the picture, with the inevitable consequence being one of distortion, imbalance, and the disintegration of culture. There is also a direct correlation between the disintegration of life systems and the decay of cultures. Albert LaChance's chapter emphasizes the reconnection of cosmology and cosmogony, drawing on Noah as the first deep ecologist. The Noah story is a model of harmony between cosmology and cosmogony with cosmogony giving form to cosmology. The firm bond between the two is the covenant, the relation between them. LaChance is well qualified to offer his solution because of his years of study and service in the area of the environment. For further insight into his views see his book *Greenspirit: Twelve Steps in Ecological Spirituality.*

John Carroll extends the meaning of love your neighbor to all other species, to the elements, and to the whole natural world. He renders explicit the meaning implied in this commandment. The whole creation is worthy of our love because God dwells within it, or it dwells and subsists in the divine. *Panentheism*—pantheism— is the proper understanding of the divine-cosmic-human connection, what Raimon Panikkar calls the *cosmotheandric* mystery.[1] Carroll then exhorts us to *live simply,* and he adds the rest of the dictum: "Live simply so others may simply live," including other species. This plea for simplicity and its implementation in our daily lives is a practical way in which each one of us may have an impact in turning our species around toward the direction of real ecological responsibility. Simplicity of lifestyle is certainly an implication of discovering and living from the depths of what Miriam MacGillis refers to as our "earthself."

William McNamara's provocative article on language pollution addresses a preliminary condition for a solution to the environmental crisis. He argues eloquently for a radical renewal or rectification of language, and observes that the healing of the planet and renewal in a general, comprehensive sense must begin with this reformation in how we use or tolerate the use of language. Frederick Levine's chapter functions on a resource level by providing a glimpse of how the land was regarded and related to in ancient Israel. This biblical, historical study suggests lessons useful to resolving this aspect of the ecological problem. David Toolan dissects the attitude of the Catholic church to the population issue, one of the more important factors to solve if the earth is to be safeguarded. Albert Fritsch concentrates on a solution to the use of technology because the ways in which society has misapplied science and technology have led to the environmental disasters we now face. He perceives a solution in a more environmentally friendly or sensitive approach, or in appropriate technology, which will then make room for the earth to heal itself.

Richard Haas applies gospel values to economics as a means to transform productivity into a more human, caring mode. He tells us that he belongs to the "loaves and fishes school of economics," a school that follows the principle of the "multiplying effect of loving relationships." He regards creativity as the deeper nature of productivity when motivated by love, and he has discerned that productivity is better understood when viewed within the context of relationships that are personal, like the models of sexuality, and the Trinity. Productivity should not be mandated but rather inspired by enlightened management, government, and a unique type of Catholic spirituality.

Mary Joyce's piece offers a more humane understanding of the view and role of the body. She flatly rejects Aristotle's definition of "man as a rational animal" and presents a more modern, integrative view that asserts the human person should be defined through the notion or experience of the bodily state or condition. This approach allows us to be at home in the world; it connects us with the created order or the cosmos, with others, and with God in a greater sense of intimacy and friendship. It also makes sense of our experience.

Marc Boucher-Colbert explores the rich connection of community, organic farming, and the spirituality that evolves from their proper interaction. He opens up another crucial part in the awakening of environmental consciousness, the critical area of food production employing farming techniques that respect the nature and character of the land. This organic method is very important, since the farming practices of the huge agribusinesses have seriously depleted the soil in much of North America, Europe, and other areas of the world. Organic farming is conducive to a spirituality of the land.

Beatrice Bruteau's "Theotokos Project" is another elaboration of panentheism but unfolded in an evolutionary context. The universe is becoming or is, through its development, the God-bearer to *Theotokos*, because it is pregnant with the divine. Bruteau's article is actually a work in a kind theological cosmology, and it is surely compatible with the story of the universe as articulated by Thomas Berry, since it essentially elaborates the metaphysical insight of the presence of the divine, the numinous reality in all of creation. The "Theotokos Project" is a variation on the sacred narrative of the cosmos.

Five of the articles in this work can be described as applied spirituality, because they act as inner resources brought to bear on changing attitudes and patterns of behavior destructive of nature. Charles Cummings presents a sacramental view of the natural world and our relation to it.[2] It is a form of natural contemplation, of "reading" the Book of Creation. Tessa Bielecki contributes insights from the mystical tradition on the stages of the spiritual life leading to transformation and mystical union with the divine. Terrence Kardong offers the wisdom of the Rule of St. Benedict toward creating positive or eco-friendly attitudes and habits. He emphasizes especially Benedict's wonderful teaching on the twelve degrees of humility, the very essence of Benedictine spirituality, stability, and frugality.

The Franciscan vision is presented by Richard Rohr, who shows that the substance of the tradition stresses a cosmic or creation-centered spirituality. His article is an application of the values of the gospel and the Franciscan assimilation of these values to the ecological crisis. The Ignatian method of spirituality is then explored by William Wood in relation to the care of the earth and the whole issue of awakening and developing awareness of the importance of the environment.

All of these approaches represent different resources that the Catholic tradition can draw upon in formulating an eco-theology and spirituality that raises the issue of the earth to the top of the church's agenda.

In what follows I want to reflect on seven areas that should also be included as elements in any solution to our planet's ongoing environmental extremity: 1) the necessity for interfaith relations; 2) the church's commitment to interreligious dialogue; 3) the Parliament of the World's Religions, the "Global Ethic," and the need of a strong moral voice for the world; 4) nonviolence as a primary value, and an answer to the increasing problem of violence; 5) spiritual resources: a) nature-mysticism and the symbolic dimension, b) the contemplative experience, and c) toward a global spirituality; 6) the visionary role of art and the imagination; and 7) the possibility of an enlightened civilization.

The Necessity for Interfaith Relations

Multiculturalism and pluralism of faith communities are becoming more and more facts of life in America, Europe, parts of Africa, Latin America, and Asia. Gone are the days when we can hide in our cultural ghettos, pretending that other traditions don't really matter. The collapse of the Berlin Wall, although it signalled the demise of a hated system of oppression, is also a graphic symbol of the age into which we are moving. In this age the barriers that have separated religions, cultures, and nations are gradually coming down, assisted by the mass media, instant communication, and the interdependence characteristic of modern life. This phenomenon is undoubtedly the cause of considerable discomfort and anxiety for some as they see themselves and their own faith community in a more vulnerable position facing the world and other traditions. It's not possible to put the genie back into the bottle, and Pandora can't close the box now that it's finally open. So, what is to be done?

The meaningful course is one of openness to the new situation and a quiet welcome of its benefits and challenges. A destructive, reactionary response would signal a retreat into a kind of fundamentalism attracted by the false psychological and intellectual security it provides, but to take that road inevitably leads to interior stagnation and potentially dangerous tensions with other faith communities. Fundamentalism is not a genuine option; it is the abrupt abandonment of the possibility of real growth.

We also need to pursue interfaith encounter, dialogue, and community in order to avoid misunderstandings; to heal old rifts; to meet around our common problems, concerns, and hopes. Interreligious dialogue and collaboration are not luxuries. They are necessary if peace is to take root, for as Hans Küng has observed (and Mahatma Gandhi before him): "There will be no peace on earth unless there is first peace among the religions."[3] Ultimately, the cause of survival requires the religious to collaborate together, and this will be more and more the case in the future. There is also a philosophical and spiritual motive behind dialogue, encounter, and collaboration that contemporary physics describes, and that is the essential unity of all reality, and so, an underlying unity among the religions as well, even though this unity is not yet explicit or clearly discernible.

The Church's Commitment to Interreligious Dialogue

The Catholic church has long been a leading force in promoting interreligious dialogue. Vatican II not only inaugurated the ecumenical movement as a mainstream concern of the church, but it set the stage for reaching out to other religious traditions as well. Pope Paul VI provided vital leadership here by creating the Secretariat for Non-Christians during the Council on May 17, 1964, on the Feast of Pentecost. The Secretariat for Non-Christians was renamed the Pontifical Council for Interreligious Dialogue in the curial reorganization of 1988. It has since become a very significant branch of the Holy See in pursuing contacts with religious and spiritual leaders in all of the other world religions. The Pontifical Council carries on consultations, organizes conferences and studies, sends delegates to major interfaith events and meetings, sponsors seminars and workshops, gathers information, and cultivates relationships with representatives of all the major traditions. It also publishes the *Bulletin*, which often carries important ecclesial documents relating to interfaith dialogue and collaborative efforts.

When still called the Secretariat for Non-Christians, this curial entity, under the authority of Cardinal Pignedoli, approached the abbot primate of the Benedictine Order—then Rembert Weakland—and asked the Benedictines, along with the Cistercians, to assume responsibility for dialogue with representatives of the Asian religions, notably with Hindus and Buddhists. The reason for this decision of Rome was that monastics cultivate a spiritual depth that has its equal in Hindu *sannyasis* (renunciates or monks) and Buddhist monks. The abbot primate gave the responsibility to the Benedictine organization in Paris called *Aide Inter-Monastique* (AIM), an agency founded earlier in the century to assist struggling monasteries in the Third World. AIM eventually established a subcommission to implement the Vatican's mandate. The subcommission is known as *Dialog Inter-Monastique* or Intermonastic Dialog (DIM). In North America a parallel entity was founded in 1978 and named the North American Board for East-West Dialogue. In 1992 the name was changed to Monastic Interreligious Dialogue (MID).

Both DIM and MID carry out similar activities of seeking contacts, organizing conferences, issuing statements, participating in important interfaith meetings, and engaging in intermonastic exchange programs involving visits to monasteries of other traditions, for example, to Tibetan Buddhist, Theravadan, and Zen monasteries in Asia. Monastics from these Oriental traditions send representatives to our monasteries in Europe and America. This exchange program has gone through many phases and is quite successful.

It also should be noted that the National Conference of Catholic Bishops has created the Committee for Ecumenical and Interreligious Affairs, and it functions somewhat like the Pontifical Council, while also advising the bishops on interfaith and ecumenical matters. The wider notion of ecumenism to include other religious traditions only came into being in the American church during the 1980s. There is also the National Association of Diocesan Ecumenical Offices (NADEO), which coordinates efforts on the more local and regional level as these efforts relate to interfaith work.

There is, furthermore, the singularly admirable achievement of Thomas Keating, a former Cistercian abbot, and erstwhile chairman of Monastic Interreligious Dialogue, who founded a group called the Snowmass Conference, named after the place where his monastery—St. Benedict's—is located. Meeting since 1984, some fifteen members, each representing a world religion and each one a spiritual master in his or her own right, have come to complete agreement on eight guidelines for interreligious understanding. These guidelines are:

1. The world religions bear witness to the experience of Ultimate Reality to which they give various names: Brahman, Allah, the Absolute, God, Great Spirit.
2. Ultimate Reality cannot be limited by any name or concept.
3. Ultimate Reality is the ground of infinite potentiality and actualization.
4. Faith is opening, accepting, and responding to Ultimate Reality. Faith in this sense precedes every belief system.
5. The potential for human wholeness—or in other frames of reference, enlightenment, salvation, transformation, blessedness, *nirvana*—is present in every human person.
6. Ultimate Reality may be experienced not only through religious practices but also through nature, art, human relationships, and service of others.
7. As long as the human condition is experienced as separate from Ultimate Reality, it is subject to ignorance and illusion, weakness and suffering.
8. Disciplined practice is essential to the spiritual life; yet spiritual attainment is not the result of one's own efforts, but the result of the experience of oneness with Ultimate Reality.[4]

It is difficult to exaggerate the breakthrough quality of these statements and the value they have in the ongoing process of interfaith encounter. They deserve to be studied, debated, reflected upon, and clarified by way of an extensive commentary that draws out their implications.

The Parliament of the World's Religions

One of the clear implications of interfaith encounter is that the problems of the world are too complex for any one tradition to handle. This is especially true of the ecological crisis. This problem is common to every religion, indeed to every sentient being. All of the other issues—the desire for peace, hunger, poverty, injustice, homelessness, violence, oppression, overpopulation, and so on—are similarly common concerns. This insight is essential to appreciate the significance of not only the Parliament of the World's Religions but the whole interfaith emergence in our time. We should keep in mind that interreligious dialogue and cooperation are among the most crucial works of the age. Through these activities the world's religions discover deep bonds of community. Out of this identity comes a powerful sense of common responsibility for the earth in all its critical needs. Dialogue leads to communion with one another, with other species, with the earth itself, and finally with the whole universe. Interreligious encounter is thus morally necessary and spiritually both vital and enriching. It becomes the basis of inner growth for the human family in a new dimension of globality totally unknown until this precious moment of history.

The Parliament of the World's Religions (held in Chicago from August 28 to September 5, 1993) was first conceived as a centennial celebration of the first Parliament of September 1893 in Chicago, where as part of the World Columbian Exposition—really a World's Fair—the religions of the planet met for the first time in history, thus setting in motion forces that would culminate in the international interfaith movement we have today. The 1893 Parliament was a major historical event just by virtue of happening, and it succeeded in capturing the imagination of the civilized world at the time, but it was not as substantive or as conscious as the 1993 Parliament.

Early in the planning process, the second Parliament evolved beyond its centennial context. It was conceived as an opportunity for a call to action on the critical issues before the international community. The second Parliament was thoroughly activist in its conception and its concrete, existential unfolding, unlike the 1893 Parliament, whose overriding concern was to promote peace in the world through mutual understanding among the traditions. Some eighty-seven hundred persons participated in the second Parliament in fifteen plenary sessions and some seven hundred workshops, seminars, dialogues, with literally thousands of presentations. The Parliament would have had thirty thousand or more participants, but it had to close registration a month before the opening because of lack of space at the Palmer House Hotel, the site of the gathering.

This second Parliament was a totally unique happening. Those assembled, representing 150 religions, the academic, arts, and professional communities, people from every possible background, came together for eight days in a beautiful celebration of the human spirit, the *anima mundi* become conscious. They poured through corridors, elevators, ballrooms, and the lobby of the Palmer House. The lobby itself became a metaphor for the Parliament in its dramatic array of color and garb, a veritable tapestry of humankind's faith expressions and cultural streams.

It was Karl Jaspers who popularized the notion of the Axial Age (the first millennium B.C.E. up to the time of Christ), when so many of the great founders and figures of the world's religions lived, like the prophets in ancient Israel; Lao-Tzu, Chuang-Tzu, and Confucius in China; Mahavira and the Buddha in India. Their insights became foundational and thus axial for the world's cultures. The 1993 Parliament, I believe, will be seen in time as an axial event-process, because it has inaugurated deep and far-reaching changes in human consciousness, the seeds of which were already apparent to all those who attended this truly watershed convocation.

The spirit or consciousness that pervaded the Parliament was one of genuine openness, real listening to one another, mutual respect, a sense of solidarity and joy, with an equal sense of urgency, expectancy, and collective responsibility. There was a palpable spirit of love and communion that permeated the gathering. It was as if the veil had been parted, and for eight days we were allowed to experience a vision of a new paradigm of relationship in which all the skills our species needs in order for our planet to survive were revealed to us. It was something that approached or suggested the Pentecost event described in Acts.

The new paradigm of relationship that became explicit during the Parliament involves an abandonment of competition and conflict between and among religions, and a shift to the realization that we belong to a universal community, the community of religions within the larger sacred community of the earth itself. This paradigm shift is, I think, axial in significance because it signals the end of millennia of religious wars based on mutual suspicion, ignorance, and hostility. There is now a genuine openness among the traditions and a willingness to learn from one another's treasury of spiritual experience. So profound was this whole occasion, this new spirit of reconciliation and collaboration, that Robert Muller, the chancellor of the UN's Peace University in Costa Rica and a former assistant secretary general of the United Nations, predicted that the Parliament of the World's Religions would ignite a spiritual renaissance all around the planet.[5]

The centerpiece of the Parliament was the 250 member Assembly of Religious and Spiritual Leaders, comprising figures from all world religions and many lesser-known ones. A number of scholars, experts, and trustees of the Parliament were part of this body. The Assembly met at the Chicago Art Institute, site of the first Parliament, in three sessions from September 2 through 4. Much of the time was taken up with the text of a statement on a new "Global Ethic," a document the Council for a Parliament of the World's Religions had prepared for the Assembly's consideration and approval. After considerable comment by a number of members, the global ethic was signed by 99 percent of the Assembly with the proviso that the title read *Towards a Global Ethic (An Initial Declaration)*. This title change reflected reservations that some of the members had with the process, content and language of the document, and so the Assembly regarded the "Global Ethic" as basically in need of further development.

Towards a Global Ethic represents a consensus on human values to guide action from the different perspectives of the various religions. It is remarkable because it is the first document of its kind that brings the faiths together and commits them to a common set of ethical principles. This interreligious ethic also incorporates the ecological dimension into the moral sphere and declares:

> We are interdependent. Each of us depends on the well-being of the whole, and so we have respect for the community of living beings, for people, animals, and plants, and for the preservation of Earth, the air, water and soil.[6]

Towards a Global Ethic ends with a commitment to "nature-friendly ways of life."[7] Millions of people have now read this document; it has become the focus of numerous conferences, and Robert Fogel, winner of the 1993 Nobel Prize for Economics and a professor at the University of Chicago, has quietly begun to teach the document to his graduate students.[8]

The Assembly also made dozens of recommendations. One of the most prominent was a near unanimous call for a permanent organization for the religions, a permanent Parliament that would act as a kind of United Nations of the religious traditions. Time ran out, so the Assembly was unable to set up the initial committee to look into this possibility. It did manage to empower the Council for a Parliament of the World's Religions to explore the feasibility of this task, come up with the most effective model of such an organization, and then inaugurate it. The effort continues, but the work is slow.

Virtually everyone in the Assembly recognized the need for a permanent structure or platform for the religions, so that religious and spiritual leaders could have a clear, strong collective moral voice in the global, regional, and local arenas. Only in this way can the leaders accept what the Dalai Lama calls our universal responsibility,[9] and so establish a new institution to serve as a global moral voice. This voice, it was felt, is desperately required in our world. It must be a voice that can never be compromised or vitiated by political considerations or pressures. For far too long the papacy has attempted to exercise this kind of function in the international arena, but it has an inherent limitation; it speaks for only 18 percent of humankind.

The religious and spiritual leaders together have the enormously difficult task of inspiring change in how people live in relation to the earth. The political leadership of the world has not succeeded in bringing about this monumental transformation to a more simple style of life. It remains for the spiritual leadership of the planet to try, with the assistance of the political, economic, scientific, philosophical, and cultural segments of society. A permanent platform for the world's faiths would greatly enhance the efforts to coordinate and apply the necessary moral leadership to each appropriate situation requiring guidance and firm encouragement. Clearly the Spirit is leading us in this direction, and it should be noted that if there is a failure peculiar to power, whether spiritual or political, it is disobedience to the Holy Spirit. There have been many grave instances of such disobedience in the church's history. Analogous failures could be adduced from the other traditions as well. A religious tradition standing alone cannot be as effective as all of them working in concert on the critical demands of the age, chief of which, in our time, is the environment.

In 1994 the Council for a Parliament of the World's Religions has created a special group called the International Interreligious Initiative from among trustees and associates of the Council. This group has the responsibility for planning and consulting with the twenty-four presidents of the Parliament, the 250 Assembly members, the two hundred sponsoring organizations, the other interfaith organizations, especially with the World Conference on Religion and Peace, the World Congress of Faiths, the Temple of Understanding, Global Forum, and so on. Out of these consultations—and additional ones with experts and other groups—a plan of action will emerge. A convening of a constitutional convention might then occur in order to draft and approve a charter. Then we will move to implementation. The process takes time, because it involves so many organizations and people; it is a huge, corporate undertaking, and we have no clear idea how it will all come out, but we do have ideas and hopes relating to the direction we would like it to go.

In this context I propose that we consider the Hindu notion of *sannyasa*—an ancient Sanskrit term for renunciation, the ideal of the monk and the name of the fourth stage of life in the Indian view of existence. *Sannyasa* denotes the way of human transcendence in search of the divine, the quest for the Absolute celebrated in India's culture, history, and religious consciousness. *Sannyasa* is really a mystical state that exists to facilitate the spiritual journey. It is also understood as a state beyond all visible religion and its structures, all social, political, and economic ties. It is beyond all, though rooted in a particular tradition, beyond all because God is found beyond everything. *Sannyasa* could be extended to members of every religious tradition, and it can lead the world into a new universal society that lives in harmony with the natural world and other species by being an example of a truly simple lifestyle— providing a zone of contemplation where *sannyasis* from all traditions can congregate. Through their example and teaching humanity could be raised to planetary consciousness.

The Primacy of Nonviolence as a Value

Any overall solution to the ecological crisis must deal with the escalating problem of violence. The degree and scope of violent behavior have reached alarming proportions, and there is no end in sight. There are some thirty wars raging on the planet. Our cities are experiencing a rising tide of violent crimes. Guns are everywhere and are used at a moment's notice with little provocation. The TV and movie industries have shown virtually no restraint in their graphic portrayal of violent acts. Psychologically, American society has an addiction to and a fascination with violence in all its forms. This addiction and fascination indicate an inner disorder and a lack of focus or spiritual centeredness in people and in the culture at large. The result is an extroverted, boisterous, and irreverent cultural milieu.

The news media has similarly lacked responsibility by constantly focusing attention on violent crimes. There now seems to be a crescendo of voices speaking out on the abuses emanating from Hollywood and TV journalism. Finally, people are realizing that the film and TV industries, along with the news media, are actually perpetuating the problem of violence by teaching it and transmitting it as a value, even though there may not be a self-conscious policy to do so.

The religions have a responsibility to respond to this social crisis. Many have done so on the level of the systematic violence of states of war, but they must do more in the area of media preoccupation with crime and terrorism, and the heinous acts associated with them. The lesson should be clear from the advertisement industry, which learned long ago the effectiveness of subliminal programming, how images have an impact on the unconscious, and how the unconscious can be manipulated, stimulating desires for products. The same psychological mechanism is at work in the entertainment media. The movie and television industries have created a need—on the unconscious level—for violence, and people seek their daily fix. The same is true of children and their attachment to horror films. What is this peculiar fixation about? It's very simple: being frightened. Children like or crave the thrill of being scared so much that they endure the appalling scenes of horror films, where adults would recoil and flee.

One key answer to a culture's preoccupation with violence is to teach, insist on, and *live* the value of nonviolence. It can be done successfully, and it has been done for more than 2,500 years by Jains and Buddhists. Neither Jainism nor Buddhism has ever supported war or personal violence; this nonviolence extends to all sentient beings. Christianity can learn something valuable from these traditions. This teaching on nonviolence has been incarnated in the lives of Mahatma Gandhi, Martin Luther King, Jr., and the Fourteenth Dalai Lama with significant results, although the victory of nonviolent struggle is still to be won in Tibet. The Dalai Lama, however, has gained tremendous moral credibility over the Chinese because—among other things—of his espousal of and commitment to nonviolence.

The Dalai Lama entered into a collaboration with Monastic Interreligious Dialogue in signing the "Universal Declaration on Nonviolence," a document that is really a declaration of religion's independence from war, violent behavior, and all governments that are warlike, that attempt to manipulate or enlist the support of religious and spiritual leaders in their campaigns of systematic violence against other nations. The "Universal Declaration on Nonviolence" was agreed to on October 7, 1990, and then signed more formally on April 2, 1991, in Santa Fe, New Mexico, during a joint press conference.[10]

The "Universal Declaration on Nonviolence" was written in a context that anticipates a new civilization, a civilization that is universal, enlightened, peaceful, and just, ecologically wise, and free of the need for armaments, since war has been outgrown as a means to settle disputes between and among states. The declaration invites us to dream of or imagine a world without war, and then to move in that direction. If we can transmit the value of nonviolence to culture on a broad scale, then the level of official violence will steadily decline, and we can begin to create a "culture of nonviolence,"[11] as the "Global Ethic" puts it, having been influenced by the earlier statement on nonviolence. As nonviolence becomes more an essential part of our global culture, the need for armies and vast armaments will greatly decrease. As we quiet down the planet and retire the whole notion of war, the attitude of nonharming will extend itself quite naturally to all other areas of human life. Gentleness will become a real possibility for humankind. The earth will then feel another pressure lifted from itself, and that will contribute to its further healing. Adopting nonviolence is one of the greatest acts of love for the planet and for one another that we can embrace, an act that will bear fruit by advancing environmental responsibility. If we want peace we must live nonviolently.

Spiritual Resources

Nature-Mysticism and the Symbolic Dimension

Nature-mysticism is a very precious form of spirituality that can be profoundly beneficial in transforming attitudes toward the earth and patterns of behavior destructive to the natural world. Eco-spirituality is actually a more contemporary type of nature-mysticism. Nature-mysticism itself is essentially panentheistic and depends on the metaphysical insight expressed—among other places—in the biblical text, "In Him we live and move and have our being" (Acts 17:28). All reality, life and being, the entire cosmos subsists in God. Nature-mysticism is first the awareness of the divine Presence encompassing us. It presupposes a sacramental understanding of the earth, the natural world, and the universe itself. It is also an experience of unity with nature. The dichotomy of the inner and outer breaks down in experiences of this type as the unitive character becomes dominant. Forrest Reid, an English writer, conveys this unitive note from his own experience:

> The whole world seemed to be within me. It was within me that the trees waved their green branches; it was within me that the skylark was singing; it was within me that the hot sun shone, and that the shade was cool.[12]

Nature-mysticism also involves the symbolic dimension of reality. Nature itself participates in this symbolic dimension that the human mind does not create but discovers. Again, this is a sacramental view of the world. It is an intuitive and experiential wisdom that is part of the *philosophia perennis* found in all the ancient cultures. It is the basis of the Franciscan vision, and the whole insight expressed in the metaphor of creation as a great book that we must learn how to read through a process of inner awakening to natural contemplation. Romantic poets in England and the Continent, American Transcendentalists, indigenous tribal societies were all in touch with this religion of nature, this mysticism of the natural world. Bede Griffiths says it best:

> We only begin to awake to reality when we realize that the material world, the world of space and time, as it appears to our senses, is nothing but a sign and a symbol of a mystery which infinitely transcends it.[13]

To see in this way, to pierce the symbol, is a form of illumination or what can be called theophanic consciousness—awareness of the divine in nature and nature in the divine. Nature-mysticism is a constitutive part of Christian spirituality and should be incorporated into the formation of an overall process of moral and spiritual education as it relates to and affects the natural world.

Contemplative Experience

Contemplative wisdom offers the church, humanity, and the earth itself the greatest single resource for changing course before it's too late. We can transform human awareness and action by allowing ourselves to be changed first. There are some 41,000 books on mystical subjects in the Christian tradition alone. Add to these those of all other traditions and it's somewhere in the neighbor-hood of 200,000 works. This is a vast storehouse of wisdom useful to the task of altering attitudes, perceptions, and patterns of living.

Contemplation is experiential. It is the gradual awakening into conscious relationship with God, and the subtle refinement of our capacity to love God and to be aware of the Divine Presence in everyone and everything. It is the capacity for other-centeredness, first in relation to the divine, and then in relation to other persons, other species, the earth itself, and the entire cosmos. This aware-ness of and relationship with the Divine Presence lead in time to active love of God and then to transforming union.

Contemplation, through an inner process of struggle and growth—the subject of the celebrated stages of purgation, illumi-nation, the dark nights, and union—refines the willing person in a radiant and exquisite sensitivity to all reality, increasing, expand-ing, and strengthening the person's ability to respond with love and compassion to each person, to each being, and to the whole created order. The basis of this expanded capacity is the diminish-ment of self-preoccupied interests, the extraordinary transformation divine love initiates and sustains in one's consciousness, charac-ter, will, and behavior. These are brought progressively under the influence of grace in the union with divine love itself. Contempla-tion grants us that skill of sensitivity that we so desperately need in order to save the earth. That gift of sensitivity is glimpsed in acts of compassion and love; it is seen in attitudes and in presence to others.

The monk Theophane relates this story: As a little boy, he vis-ited the magic monastery. Here he was deeply inspired by the humility of an old monk who had given a tour of the monastery to a group of guests. He noticed that the monk never talked about himself. So the little boy, who is actually Theophane, asks the old man to say something about himself. After a long silence, the ven-erable monk said: "My name . . . used to be . . . Me. But now . . . it's you."[14] This story illustrates well the fruit of radical transfor-mation that contemplative interiority facilitates.

Our tradition has vast wisdom in this area, with ample material for practical guidance, but I know of no greater spiritual master in the Christian tradition in our time than Abbot Thomas Keating.[15] Abbot Thomas has developed a practical mystical teaching on con-templative prayer that he calls the Divine Therapy. The human condition is one of profound illness. The doctrine of original sin names this illness in a sense. The illness can be characterized as self-centeredness, which blinds us to everything: to God, to others, to our true self as an image of divine love and compassion, to the natural world, and to the reality of life. Blindness, in this sense, is in the system of the false self, that system which relates to the instinctual needs of esteem, security, and control, which the false self is always seeking to fulfill. The suppression of emotional pain that occurs when these desires are frustrated, and the compensation contrived to deal with the pain are part of the labor of the false self and its elaborate system for happiness through its program of seeking gratification of these instinctual needs or compulsions.

Centering prayer, as a contemplative method, is part of the Divine Therapy that gradually unmasks the false self; reveals the hidden springs of motivation and distinguishes among desires for security, esteem, and control; and leads the person into real security—reposing in the Presence of God. Centering prayer is a way to become inwardly still so that we may be conscious of the Divine Presence. As we grow in our awareness of this Presence and are healed and freed from the false self, we also grow in our capacity to love and feel genuine compassion for all beings. The Divine Therapy brings us deeper into contemplation and evolves our capacity to respond spontaneously in loving, kind, and compassionate ways.[16]

Only when we are drawn into divine union or the transforming union with God—the revolutionary summit—can we say that we are fully cured of the roots of sin and illness. Union is the goal of the Divine Therapy, and we move progressively toward it in the practice of Centering prayer, which is the heart of the Divine Therapy itself. In the practice of the prayer, a profoundly Christian form of non-discursive meditation, or imageless, wordless, and "thoughtless" prayer, we move through four "moments" in a circular motion each time we sit. We begin with the sacred word as vehicle of our intention to be present and to give ourselves to God. We can use any word we like—*Jesus, Abba, Mother, love, peace, forgiveness*—beginning always with the sacred word and maintaining focus on it during the time of prayer is the first moment.

The second moment is a period of rest in which we have a sense of God's Presence, and we experience deeper and deeper degrees of interior silence. Then we are drawn into the third moment, a time of unloading, where as a consequence of deep rest of our body, mind, and spirit, raw, undigested emotional wounds or contents from early childhood emerge. This can be extremely disturbing and painful, but it is part of the process of Divine Therapy the Spirit is taking us through. This is what the dark nights accomplish in us: the purification of our senses and the will. Unloading then leads into the fourth moment, that of evacuation of primitive or raw thoughts and emotions. We then return to the sacred word and start again.[17] Such is the mechanism of the Divine Therapy and the work the Spirit does in us in contemplative prayer.

Toward a Global Spirituality

For some years now the conviction has been growing in me that spirituality is ultimately what unites humankind; indeed, it is spirituality that is the true religion of the human family. I don't mean to imply a super form of spirituality subsuming all others. Rather I refer to the mystical core experience of each authentic religious tradition, the aspect that can be regarded as the mature expression of faith when enlightened by contemplative experience and wisdom. There are many kinds of spirituality, but all have in common the dimension of spirituality itself, and this realm of depth is itself constitutive of being human. Now this conviction was only confirmed for me at the Parliament of the World's Religions, since the thousands who attended were not seeking a religion, but rather, a spiritual path or spirituality.

This question naturally suggests itself: Can there be a universal or global expression of spirituality? There is the global ethic, a consensus among the traditions on ethical values and guidelines. Is it possible to discover a similar consensus on spiritual values, or are the obstacles too insurmountable because the religions are so different? I tend to think some sort of global spirituality is possible on the basis of the "Guidelines for Interreligious Understanding" of the Snowmass Conference. Just as there can be agreement on certain theological principles held in common by the world religions, and just as they can also agree to the norms of the global ethic, the religions can certainly find a kind of consensus on spirituality. The world religions could, perhaps, agree to the spiritual utility of some sort of meditation practice and a discipline of spiritual reading, also fasting, service to the community, chanting, self-denial or asceticism, a program of study, and so forth. Perhaps a group should get together and work on the feasibility of a universal spirituality. I doubt that it can be simply concocted or thrown together. The inner, hidden essence of spirituality consists in that radical openness to God or Ultimate Reality that occurs when we stand in the Presence in the depths of our being in our utter nakedness and poverty. That act of openness and trust, of reaching out to God in faith, *that* is spirituality. We must begin with that common experience in creating a planetary spirituality.

Sacred Seeing: The Visionary Role of Art and the Imagination

Art and imagination offer further resources of the human spirit open to the earth, the cosmos, and the divine. Art, like poetry, is a language that transcends the limits of the rational, analytic mind. It is able to leap beyond the self-imposed boundaries of conventional knowledge and understanding. It is able to see and intuit relationships hitherto unsuspected. It can reveal and unlock reservoirs of feeling and insight not perceived by the conscious mind. It is able to express what most hardly dare to think, and it can see the future to some extent by predicting the outcome in the clash of ideology with truth; for example, the visionary art of the Russian painter Nicholas Roerich, whose many pieces often portrayed the theme of the persecution of the Russian Orthodox Church by the Soviet regime. One can discern in his art with this focus the intuition of the church's ultimate triumph.

All forms of art can be enlisted to communicate the values of eco-spirituality. Through the use of artistic expression and the formation it makes possible, the young and future generation can be educated in a right attitude toward the earth, the elements, and other species, all within the context of the Christian faith in dialogue with the other traditions and the resources of humankind's spiritual treasures. Most of all, art can teach that the earth is the locus of all value and the matrix of all our relationships, the context of our own individual spiritual journeys. Art can lead us to God through its reflection of beauty, which itself gives us an indication of God's reality. We can find God in art through the encounter with beauty and the mysterious art work of the divine Artist, the cosmos. God is more of an artist than a philosopher, because God portrays or paints reality in the creation, like an artist rendering a work on a canvas. God gives us a picture of truth in the natural order, and this is again the whole symbolic realm.

Similarly, the imagination, the inner eye of artistic creation, is able to glimpse things, states, situations, ideas, beings, and places ordinarily inaccessible to conceptuality. Consider how many scientific discoveries and developments were once only in the domain of science fiction. The human imagination is a sacred faculty for envisioning and extending the reach of our knowledge. Through the creative imagination we are able to feel and dream our way into the future. The divine can speak to us, and does speak to us through the gift of the imagination, as well as the dream state itself. The imagination should be applied to the work of altering cultural perspectives, creating new vistas toward which humanity must strive.

The Possibility of an Enlightened Civilization

The great task of the next millennium—commencing in our time—is the evolution of a universal civilization predicated on ecological awareness and responsibility, a global society and culture animated by compassion, kindness, love, and humanity, living in a sustainable and harmonious relationship with the earth. It will be a wise, just, peaceful order, where hunger, poverty, homelessness, and disease will be no more. Its citizens will possess a spiritual vitality, a moral clarity, a psychological and emotional integrity, because they are one with the center of their being, the Divine Reality to which they will be united in the bonds of mutual love. This new global order will be a civilization with a heart. Humankind will have come of age and will anticipate the kingdom of heaven. All of our utopian dreams, recorded in many of our myths, will be realized when this new spiritual society dawns.

The Parliament of the World's Religions, or some variant of it—including Catholic spiritual traditions—will become the vehicle for our planet to achieve this universal civilization with a heart, a global culture in which spirituality is the most treasured value, a spirituality that unites earth with heaven and cherishes the natural world and the cosmos as sacraments of the Divine Presence. We will dwell within the Presence in the stillness, *that* stillness, the awe of creation in communion with the divine.

Notes

1. See Raimon Panikkar, *The Cosmotheandric Experience: Emerging Religious Consciousness* (Maryknoll, N.Y.: Orbis Books, 1993).

2. See Charles Cummings, *Eco-Spirituality: Toward a Reverent Life* (Mahwah, N.J.: Paulist Press, 1991).

3. Hans Küng, "Presentation: Reflections on the 1993 Parliament of the World's Religions," an Address to the Parliament of the World's Religions. Available from Teach'em (cassettes), 160 East Illinois St., Chicago, IL 60611 (800-225-3775).

4. Thomas Keating, "An Experience of Interreligious Dialogue," in *A SourceBook for the Community of Religions* (Chicago: Council for a Parliament of the World's Religions, 1993), pp. 106-7. See also *Speaking of Silence*, ed. Susan Walker (Mahwah, N.J.: Paulist Press, 1987), pp. 126-29. Guideline 8 is different from the one in the *SourceBook*.

5. Robert Muller, "Interfaith Harmony and Understanding," an address to the Parliament of the World's Religions. Available from Teach'em (see note 3).

6. *Towards a Global Ethic (An Initial Declaration).* 1993 Parliament of the World's Religions (Chicago: Council for a Parliament of the World's Religions, 1993), p. 1. For the complete text with history and commentary, see *A SourceBook for the Community of Religions*, rev. ed., ed. Joel Beversluis (Chicago: Council for a Parliament, 1994). Found also in *The SourceBook Project* (available from 1039 Calvin SE, Grand Rapids, MI 49506).

7. Ibid., p. 2. See also Hans Küng and Karl-Josef Kuschel, eds., *A Global Ethic: The Declaration of the Parliament of the World's Religions* (New York: Continuum, 1993). This contains an extensive commentary on the

"Global Ethic" and a good deal of historical detail on the formation of the document and the Parliament.

8. See Richard Longworth, "Morality: It's Time for a Closer Look," *Chicago Tribune*, *Perspectives* section, Sunday, 17 October 1993, p. 1.

9. Tenzen Gyatso, the Fourteenth Dalai Lama, *The Global Community and the Need for Universal Responsibility* (Boston: Wisdom Publications, 1990).

10. For the text of the "Universal Declaration on Nonviolence," see *SourceBook*, p. 199.

11. "Global Ethic," p. 2 (Council for a Parliament edition).

12. Forrest Reid, *Following Darkness* (London, 1902), p. 42.

13. Bede Griffiths, *The Golden String: An Autobiography*, rev. ed. (Springfield, Ill.: Templegate, 1980), p. 181. For an excellent treatment of nature as symbolic of God, see Syyed Hossein Nasr, *Knowledge and the Sacred*, (New York: Crossroad, 1981), pp. 189-220.

14. Theophane the Monk, *Tales of a Magic Monastery* (New York: Crossroad, 1981), p. 18.

15. See especially Thomas Keating, *Open Mind, Open Heart: The Contemplative Dimension of the Gospel* (Rockport, Mass.: Element Books, 1986 and 1992); *Invitation to Love: The Way of Christian Contemplation* (Rockport, Mass.: Element Books, 1992); and *Intimacy with God* (New York: Crossroad, 1994). For information on his movement, write to Contemplative Outreach, 9 William Street, P.O. Box 737, Butler, NJ 07405 (201-838-3384).

16. See Keating, *Intimacy with God*, pp. 72-91.

17. Ibid. p. 77.

A Prayer for Animals

Hear our prayer, O God, for our friends, the animals who are suffering. For all that are overworked and underfed and cruelly treated, for all wistful creatures in captivity that beat against their bars. For any that are hunted or lost, deserted or frightened or hungry, for all that are in pain or dying and for all that must be put to death. We entreat for them all thy mercy and pity, and for those who deal with them, we ask for a heart of compassion and gentle hands and kindly words. Make us true friends to animals and to share a blessing of the merciful for the sake of the tender-hearted Jesus Christ Our Lord. Amen.

This prayer hangs framed in the Church of St. Michael and All Angels in the village of Berwick, East Sussex, in the Diocese of Chichester, England.

Contributors

Thomas Berry (Passionist) is internationally known as an historian of religion and culture, and in recent years as a geologian. An expert in the ramifications of the thought of Teilhard de Chardin, Father Berry is the author of two immensely influential recent books, *The Dream of the Earth* and (with Brian Swimme) *The Universe Story*.

Tessa Bielecki (Carmelite) is co-founder and abbess of the Spiritual Life Institute, a small Roman Catholic ecumenical monastic community of men and women in Colorado and in Nova Scotia who embrace a vowed life of solitude in the spirit of St. Teresa of Avila and St. John of the Cross. Her central concern is the creation of whole, healthy human environments, and the community she founded is marked by earthy mysticism and Christian humanism. She is author of *Teresa of Avila: Mystical Writings* and *Holy Daring*, both published in 1994.

Marc Boucher-Colbert is a farmer in Portland, Oregon. He co-founded Urban Bounty Farm, a community-supported, organic farm. He has a degree in religious studies from the Catholic University of America. After a short time teaching high school religion in New Hampshire, he apprenticed under farmer Trauger Groh, a leader in community-supported agriculture in the United States. He lives with his wife, Jeanine, and tries to bring about a greater harmony between Catholicism and agriculture.

Beatrice Bruteau is founder of the *Schola Contemplationis* ("School of Contemplation"), an international network community for contemplatives of all traditions. Rooted in a religious background that comprises both Vedanta and Catholicism, Bruteau is author of *Evolution Toward Divinity* on Teilhard de Chardin and most recently of *Radical Optimism*.

John E. Carroll is Professor of Environmental Conservation at the University of New Hampshire. He is known for two books on international dimensions of environmental questions: *International Environmental Diplomacy* and *Acid Rain: An Issue in Canadian-American Relations*. He has been both Visiting Professor of Human Ecology at The University of Edinburgh and a Kellogg Foundation National Fellow. He has served as a Fellow of the Monterey Institute for International Studies and is interested in bridging the gap between religious and scientific understandings of the environment.

Charles Cummings (Cistercian) has been a monk of Holy Trinity Abbey, Huntsville, Utah, since 1960. There the monks attempt to earn their

277

living from the land. Father Charles is employed in food service for the retreatants and is on the novitiate formation team. He holds a degree in formative spirituality from Duquesne and is the author of *Eco-Spirituality: Toward a Reverent Life*.

Albert J. Fritsch (Jesuit), a self-confessed Kentucky farm boy, is an organic chemist and lover of "simple things." He is also director of Appalachia—Science in the Public Interest and a member of the Directors of the National Catholic Rural Life Conference, and advises the Center for the Study of Commercialism in Washington, D.C., as well as the Resource Policy Institute in Los Angeles and the Iowa-based Catholic Rural Life Conference. Father Fritsch is the author of *Down to Earth Spirituality* (1992); *Waste Minimalizing: Widening the Perspectives* (1994); and *Earth Healing* (1994).

Paula González (Sister of Charity) is a futurist, educator, and biologist. A former professor of biology at Mt. St. Joseph College in Cincinnati, she is now engaged full-time in the work of raising consciousness about healing the earth. Sister Paula and her companions live on the grounds of their motherhouse in a chicken barn that has been converted into a passive solar house.

Richard C. Haas is business manager of *Commonweal* magazine in New York City and president of Readership Projects of Sea Girt, New Jersey, a consulting firm specializing in services for Catholic publications. He is the editor of *Mission, Marketing & Management: Development Guide for Diocesan Newspapers*, published by the Catholic Press Association. He writes "The Market Place" column for *Living Prayer* magazine and is the author of the forthcoming *We All Have a Share: A Catholic's Vision of Prosperity Through Productivity* (ACTA Publications, Chicago, 1995), from which this essay is adapted.

Mary Rosera Joyce of St. Cloud, Minnesota, is a writer and former teacher of philosophy, psychology, sociology, and literature. She did doctoral studies in philosophy at St. Louis University and is the author of nine books and numerous articles on human life, friendship, and sexuality. Two recent books are *Friends for Teens* (1990), and *Friends with God* (1994).

Terrence G. Kardong (Benedictine) is a monk of Assumption Abbey, Richardton, North Dakota. He has graduate degrees from Catholic University and Rome. Editor of *The American Benedictine Review*, Father Terrence has written seven books and forty articles, mostly on monastic topics.

Albert J. LaChance is married and the father of two daughters. A student of Thomas Berry, he is a member of the adjunct faculties of the University of New Hampshire and the College for Lifelong Learning. He owns and operates LaChance Counseling Services and is the founding director of The Greenspirit Institute. He is the author of *Greenspirit:*

Twelve Steps in Ecological Spirituality (Element Books, 1991) and is working on a book entitled *The Architecture of the Soul: Numenal Process Psychology.* Also unpublished is his 1400-line poetic prophecy *Jonah.*

Frederick G. Levine holds a master's in theological studies from Harvard Divinity School and is enrolled as a doctoral candidate at Graduate Theological Union, Berkeley, where he is specializing in Jewish mysticism and cultural-historical studies of Judaism. He is the author of numerous articles on spirituality, comparative religions, and alternative medicine.

Miriam Therese MacGillis (Dominican) is director of Genesis Farm, Blairstown, New Jersey, which she co-founded in 1980 with the sponsorship of the Dominican congregation. With a background as a teacher, an archdiocesan peace-and-justice educator for Newark, and a program coordinator for Global Education Associates, Sister Miriam now lives and works at Genesis Farm, which she describes as a learning center for people of goodwill to search for ways to live in harmony with the natural world and each other. She also lectures widely on eco-spiritual matters.

William McNamara (Carmelite) has been a Discalced Carmelite since 1944 and is co-founder of the Spiritual Life Institute, a small monastic community of men and women, Roman Catholic in origin, universal in outreach. Father McNamara is author of many books and articles; in his prime he traveled widely as a retreat leader and speaker, giving conferences in every state in the United States but Alaska.

Richard Rohr (Franciscan) is founder and animator of the Center for Action and Contemplation in Albuquerque, New Mexico. He spent fourteen years as pastor of the New Jerusalem Community, gathered in a working-class neighborhood of Cincinnati. Father Rohr is well known for his audio and video cassettes, books that have been made from them, and for his desire to unite the work of religion, psychology, and social justice.

David M. Sherman, also known as His Holiness Bhakti Ananda Goswami, is a Catholic hermit, Triunda Sannyasa (renunciate), and Shiskha (instructing) spiritual master in the ancient Brahma-Madhva-Gaudiya lineage of Vaishnavism. His life is devoted to exalting Jesus Christ as Glory of Israel and Light of All Nations.

Wayne Teasdale is a member of a group of Benedictine-inspired contemplatives. A student and fellow pilgrim with the late Bede Griffiths in the search to bring about deeper understanding between Christians and members of Asian, chiefly Indian, religious traditions, Brother Wayne is a member of Monastic Interreligious Dialogue and vigorous participant in the 1993 Parliament of World Religions and in bodies dedicated to following up the Parliament's proposals and actions.

David S. Toolan (Jesuit) is associate editor of *America* magazine. Known for his penetrating analysis of the intersection of spiritual, theological, and scientific issues in contemporary America, Father Toolan's *Facing West from California's Shores: A Jesuit's Journey into New Age Consciousness* is a premier study of consciousness issues in contemporary American religious and spiritual movements.

Keith Warner (Franciscan) is a geographer and natural historian based in California. For nine years a member of an ecumenical lay community, he has served in a variety of multicultural ministries, among which was planting over 600,000 trees in the Pacific Northwest while working for its reforestation cooperative. After studies in geography and environmental studies he joined the St. Barbara Province of Franciscan Friars.

William J. Wood (Jesuit) directs Santa Clara University's Eastside Project, a course-imbedded program of education-in-action that nurtures a partnership of solidarity between the university and the marginalized. He has worked for twenty years probing the cause of hunger in a world of plenty, a quest that led him to the eco-justice and eco-spirituality movements.